SpringerBriefs in Applied Sciences and Technology

SpringerBriefs present concise summaries of cutting-edge research and practical applications across a wide spectrum of fields. Featuring compact volumes of 50 to 125 pages, the series covers a range of content from professional to academic.

Typical publications can be:

- A timely report of state-of-the art methods
- An introduction to or a manual for the application of mathematical or computer techniques
- A bridge between new research results, as published in journal articles
- A snapshot of a hot or emerging topic
- An in-depth case study
- A presentation of core concepts that students must understand in order to make independent contributions

SpringerBriefs are characterized by fast, global electronic dissemination, standard publishing contracts, standardized manuscript preparation and formatting guidelines, and expedited production schedules.

On the one hand, **SpringerBriefs in Applied Sciences and Technology** are devoted to the publication of fundamentals and applications within the different classical engineering disciplines as well as in interdisciplinary fields that recently emerged between these areas. On the other hand, as the boundary separating fundamental research and applied technology is more and more dissolving, this series is particularly open to trans-disciplinary topics between fundamental science and engineering.

Indexed by EI-Compendex, SCOPUS and Springerlink.

Ahmad Fitriadhy ·
Sayyid Zainal Abidin Syed Ahmad ·
Azman Ismail · Fatin Nur Zulkipli ·
Andreas Öchsner
Editors

Advancing Interdisciplinary Knowledge and Innovation

In Marine Technology

Springer

Editors
Ahmad Fitriadhy
Program of Naval Architecture
Universiti Malaysia Terengganu
Kuala Terengganu, Malaysia

Sayyid Zainal Abidin Syed Ahmad
Maritime Technology and Naval Architecture
Universiti Malaysia Terengganu
Kuala Terengganu, Malaysia

Azman Ismail
Maritime Engineering Technology and Centre for Women Advancement and Leadership, Malaysian Institute of Marine Engineering Technology
Universiti Kuala Lumpur
Lumut, Perak, Malaysia

Fatin Nur Zulkipli
College of Computing, Informatics and Mathematics
Universiti Teknologi MARA
Machang, Kelantan, Malaysia

Andreas Öchsner
Faculty of Mechanical Engineering
Esslingen University of Applied Science
Esslingen, Baden-Württemberg, Germany

ISSN 2191-530X ISSN 2191-5318 (electronic)
SpringerBriefs in Applied Sciences and Technology
ISBN 978-3-032-01308-8 ISBN 978-3-032-01309-5 (eBook)
https://doi.org/10.1007/978-3-032-01309-5

This Springer imprint is published by the registered company Springer Nature Switzerland AG
The registered company address is: Gewerbestrasse 11, 6330 Cham, Switzerland

Preface

These research papers span diverse disciplines, aiming to advance knowledge and innovation. They begin with a focus on maritime safety, introducing a long-range communication device for recreational boaters, enhancing safety and connectivity. Transitioning to transportation, researchers explore innovations like portable recycled humps on ferries to prevent roll-off incidents, ensuring passenger safety. The financial landscape evolves with machine learning-optimized credit scoring for pay-later services, promoting financial inclusivity. Studies on the awareness in regards to the convention on the international regulations for preventing collisions at sea among fishermen contribute to maritime safety. The papers extend to risk assessment, urban planning, transportation, and maritime education, enhancing safety, efficiency, and knowledge across domains. These papers represent a multidisciplinary journey, reflecting the pursuit of knowledge and innovation to create a safer, interconnected, and sustainable world.

Kuala Terengganu, Malaysia	Ahmad Fitriadhy
Kuala Terengganu, Malaysia	Sayyid Zainal Abidin Syed Ahmad
Lumut, Malaysia	Azman Ismail
Machang, Malaysia	Fatin Nur Zulkipli
Esslingen, Germany	Andreas Öchsner

Contents

Satellite Communication Device for Recreational Boaters in Malaysia Waters 1
Ahmad Faizal Ahmad Fuad, Muhammad Amsyar Mohd Anif, Amir Syawal Kamis, Mohammed Ismail Russtam Suhrab, and Muhammad Suffi Mohd Ramli

Acknowledging the Purpose of Adding Temporary Portable Recycled Hump to Ferries that mitigate the Effects of Roll-Off and Lost-Control Incidents on Passenger Automobiles 9
Sri Murugan Saga Devan, Noor Apandi Osnin, Bhisham Singh Bhuller, Sofia Irdina Shamiri, and Ain Nabila Mohamad Khalil

A Credit Score in Pay Later Service by Using Machine Learning Optimization: A Bibliometric Analysis 17
Yosy Arisandy, Yosza Dasri, and Shahrul Nizam Salahudin

Awareness of COLREGS Among Fishermen in Malaysia 33
Silloshina Manoharan and Mohammed Ismail Rustam Suhrab

Risk Assessment Study on Ship Accidents at Port of Tanjung Pelepas 39
Nurul Najwa Nursyafrinna Mahzil, Muhamad Nasir Rahmatdin, and Mohammed Ismail Russtam Suhrab

Designing Quantitative and Qualitative Assessment for Walkability and Bikeability Within the Railway Environment in Greater Jakarta 47
Widyanesti Liritantri, Widyarini Weningtyas, Kardina Nawassa Setyo Ayuningtyas, and Anisah Pratiwi

Study of Ship Handling in Wheel and Steering Commands for Simulator and Onboard Ship Training of Nautical Students 55
Rasyidah Mohd Sani, Mohammed Ismail Russtam Suhrab, and Nik Muhamad Afiz Harom Junoh

Enhancing Safety: Investing the Impact of Support Services on Accident Reduction in Chittagong Port and Outer Anchorage of Bangladesh 63
Tamim Ahmed, Mohammed Ismail Russtam Suhrab, Ahmad Faizal Ahmad Fuad, Noor Apandi Osnin, and Kasypi Mokhtar

A Study on the Establishment of Vessel Traffic Services at Penang Port 71
Ahmad Faizal Ahmad Fuad, Nur Iman Insyirah Zamri, Amir Syawal Kamis, and Mohd Saiful Izwaan Saadon

Development of Smart Mushroom Farming System Using IoT 79
Muhammad Ilyas Zulkhairi, Nurul Atiqah Hasman, Izanoordina Ahmad, Nor Hidayah Abdul Kahar, Siti Marwangi Mohamad Maharum, and Zuhanis Mansor

Evaluation of Implementation of Occupational Health and Safety Procedures for Airport Operations at PT Angkasa Pura I Sultan Hasanuddin International Airport Post COVID-19 Pandemic 89
Akbar Fadillah, Muhammad Rafly Ananta, Datunabolon, and Retno Mawarti S. Sos

Critical Success Factors for Improvement Initiative Deployment 95
Faris Hakim Mazlan and Mohamad Ikbar Abdul Wahab

Integration of Quality and Technology Acceptance Model (Q-TAM) to Evaluate Online Teaching and Learning Medium 103
Noor Syafiqah Redzwan, Sallaudin Hassan, Mohd Haziq Zakaria, Mohd Aliff Afira Sani, Jimisiah Jaafar, Rahmat Roslan Buang, Mohd Farid Shamsudin, and Waqia B. Fokeena

The Influence of Green Port Practices on Corporate Image: A Malaysian Perspective 111
Amirul Husaini Ibrahim, Amayrol Zakaria, Aminuddin Md. Arof, and Shamila Azman

Challenges Faced by Seaports in Adoption of Industrial Revolution 4.0: A Review 119
Kaannappar Ponniah Chandrasekaran, Amayrol Zakaria, and Aminuddin Md. Arof

Satellite Communication Device for Recreational Boaters in Malaysia Waters

Ahmad Faizal Ahmad Fuad, Muhammad Amsyar Mohd Anif, Amir Syawal Kamis, Mohammed Ismail Russtam Suhrab, and Muhammad Suffi Mohd Ramli

1 Introduction

According to Britannica Dictionary [1], boating is the activity of traveling or going by boat for pleasure. Types of boats used are powerboats and sailboats. A powerboat is a type of boat moved by using an engine, while a sailboat is a type of boat moved by using a sail.

The registered numbers of recreational boats in the USA from 2016 to 2020 are 11,861,811; 11,961,568; 11,852,969; 11,878,542; and 11,838,188 respectively [2–4]. These large numbers over the years contribute to boating accidents. The number of accidents according to type from 2018 to 2020 is shown in Table 1.

In Malaysia, the Langkawi International Yacht Registry (LIYR) is the official yacht registry in Malaysia for sailing and motor vessels. The registration is open to individuals or companies. The total number of LIYR registrics till 2018 was 1,146 vessels [5]. The registry does not issue any accident statistics.

A. F. Ahmad Fuad (✉) · M. A. Mohd Anif · A. S. Kamis · M. I. R. Suhrab
Nautical Science and Maritime Transportation Program, Faculty of Maritime Studies, Universiti Malaysia Terengganu, Kuala Nerus, Malaysia
e-mail: faizalfuad75@gmail.com

M. A. Mohd Anif
e-mail: amsyar.anif12@gmail.com

A. S. Kamis
e-mail: amirsyawal87@gmail.com

M. I. R. Suhrab
e-mail: m.ismail@umt.edu.my

M. S. M. Ramli
Malaysia Maritime Enforcement Agency, Penang, Malaysia
e-mail: muhdsuffi@mmea.gov.my

A. Fitriadhy et al. (eds.), *Advancing Interdisciplinary Knowledge and Innovation*, SpringerBriefs in Applied Sciences and Technology,
https://doi.org/10.1007/978-3-032-01309-5_1

Table 1 Number of accidents according to type from 2018 to 2020

Accident type	Number of Accidents by year		
	2018	2019	2020
Collision with recreational vessel	1028	1071	1379
Flooding/Swamping	443	399	589
Grounding	367	413	484
Falls overboard	274	299	335

Source [2–4]

Safety regulation related to local recreational boats depends on local authority and not any international regulations. Safety of commercial vessels engaged in international voyage is determined by the Convention on the Safety of Life at Sea 1974 or SOLAS 1974 developed by the International Maritime Organization [6]. The convention is a prescriptive regulation and is divided into 12 chapters that cover from safety construction, life-saving appliances, radiocommunications, safety management, and maritime security.

This study is focusing on communication devices for recreational boaters in Malaysia. Risk of accidents and fatalities among recreational boaters is inevitable even when they operate near shore or offshore. Most of recreational boaters have a limited financial capability; thus most of the money is spend on boat's rental, boat's fuel, and fishing gear. This would limit their capability to procure advanced safety equipment for recreational boating. Basic safety equipment that they carry are life jackets and mobile phones. The operation of mobile phones is limited within the coverage of the mobile phone cellular network on land, and offshore boating is beyond the coverage of the network. Satellite communication would overcome this communication area gap by providing unlimited network coverage on the sea area. However, it is costly for them to procure the satellite communication device and maintain its service. Even worse, certain recreational boaters do not know the existence of the mobile satellite communication device. Therefore, the objectives of this study are as follows:

a. To identify the current marine communication devices used by recreational boaters in Malaysia.
b. To identify the factors used for the selection of a marine communication device for recreational boaters in Malaysia.
c. To propose the most suitable communication device for recreational boaters in Malaysia based on the identified factors.

2 Literature Review

The carriage requirement of marine radiocommunication for commercial vessels engaged in international voyage is stipulated in Chap. 3 SOLAS 1974 [7]. The carriage requirement of radiocommunication is according to the type and sea area. Sea Area 1 is an area covered by at least one very high-frequency (VHF) coast radio station [8]. Sea Area 2 is an area covered by at least one medium frequency (MF) radio coast station. Sea Area 3 is an area covered by at least one communication satellite such as Inmarsat.

Marine VHF radio has dedicated frequency and channel for marine use. The radio uses a frequency range from 156.0 to 162.025 MHz, and a total of 65 channels are allocated for marine VHF radios [9]. The operational range of a VHF marine radio depends on the power of transmission and the height of antenna. A VHF radio is intended for short range communication, namely from 8 to 16 km [10]. Therefore, a marine VHF radio is not able to contact shore authorities for request of assistance. The setup of a marine VHF radio is simple, and most of the yachts are fitted with this radio as a basic communication device.

A marine MF radio has a range greater than the marine VHF radio. The range of the marine VHF radio is 320 km with transmission power of 150 W [8]. With the given range, a marine MF radio allows vessels offshore to contact the authority at shore. The setup of a MF radio is more complicated than for the marine VHF radio; it requires far more power to operate, and the costs are significantly higher than for VHF radio.

Satellite communication system is comprised of three segments, namely mobile earth station, space segment, and ground segment [11]. The GMDSS compliant Inmarsat satellite system is relatively a complicated setup, requires significant power to operate, and significantly costs higher than for the marine VHF radio. Thus, the system is not suitable to be fitted on small recreational vessels.

The U.S. Department of Justice has released a Guide for the Selection of Communication Equipment for Emergency First Responder in 2002. The recommended factors are maximum transmitter output power, secure communications compatibility, programmability, user capability, line of sight, power requirements, battery life, battery locking ability, vehicular adapter (portable radios), digital communications compatibility, durability, unit cost, operator skill requirements, and training requirements [12].

3 Methodology

The overall research process was conducted in four steps. The first step is to conduct a literature review on portable satellite communication equipment. The literature search was conducted using the website of Google Scholar, namely scholar.google.com. Google Scholar is a powerful database for scholarly publication

[13]. Then keywords were typed in the search bar, namely mobile satellite communication device. The time range was set from 2010 to 2021. The search has identified 49,600 relevant articles. The information on each article was the title of the article, name of authors, and name of the publication of the article. To determine whether the articles were relevant to the research, the title of each article was checked. If relevant, the article was clicked to see more information and the abstract. If the information in the abstract was relevant to the research, then the full article was downloaded and read.

The second step is to develop the questionnaire that is answering the research's objective. The questions would obtain information of the communication device used by the boat's owner, efficiency of the device in providing satellite communication service, and cost of procurement and operating the device. The standard response for the survey is using a 6-point Likert's scale, from 1 strongly disagree to 6 strongly disagree [14].

The third step is to conduct the survey. The survey was conducted at two marinas, namely Duyong Island Marina (DIM) located in east coast Peninsular Malaysia and Telaga Harbour Marina (THM) located in west coast Peninsular Malaysia. DIM is the largest marina in the east coast of peninsular Malaysia with less than 100 berths, while THM is one the main marinas in the west coast of Peninsular Malaysia with less than 100 berths [15]. A convenience sampling method was used for the survey. The reason for this method was used because the berths were an off-limit area to other than boat's owner and marina's staff. Therefore, the chances to meet the owner is at the jetty access gate and marina's restaurant.

The fourth step is to analyze the data collected from the survey. The data was analyzed using Microsoft Excel to determine descriptive statistic of the data, namely maximum, minimum, mean, variance, and frequency. Results obtained from data analysis would be used to answer the research objectives.

4 Results and Discussion

The survey was conducted in the first week of December 2021. During the time, the movement control order (MCO) was still enforced. In the Standard Operational 3 of the National Rehabilitation Plan, interstate movement was allowed only to individuals that had completed at least two doses of vaccine. Number of foreigners in Malaysia during the time was significantly small compared to the period before MCO started from March 2020. Due to that, the number of respondents that was interviewed was 17.

Three types of communication devices were used by the respondents, namely satellite messaging device, marine MF/HF radio, and marine VHF radio. The most popular was the marine VHF radio with 17 records, while the MF/HF radio and satellite messaging device were not popular with 1 record each. This result answered the first objective and shows that boaters are only willing to invest in basic communication equipment at sea, namely marine VHF radio and mobile telephone. None of these communication devices would cover beyond the coverage of mobile cellular network and marine VHF radio. This also shows that the awareness of Malaysian boaters on safety is considerably low when it comes to communication device and distress call.

The second objective was to identify the factors used for the selection of marine communication devices for recreational boaters in Malaysia. The factors were identified by a combination of literature review and expert's opinion. The identified factors were coverage, two-way communication, distress signal, portability, waterproof, and unit cost (price). Another factor added by experts was the subscription fee. The rating for each factor was calculated by using the average rating value (ARV). Results of rating of factors were arranged from high to low, namely portability (18.04%), subscription fee (17.37%), two-way communication (16.5%), coverage factor (14.5%), distress signal (13.46%), unit cost or price (12.42%), and waterproof (7.71%). The highest rated factor was the portability. This feature is very important to boaters because they need a portable communication device to carry to their boat during a trip and bring back home after the trip for safe keeping. Portability also allowed boaters to carry the equipment anywhere on their boat. The second highest rated factor was subscription fee. This factor is important whether the operator can maintain the service provided by the telecommunication company according to operator financial capability. The third highest rated factor was two-way communication. This feature enables boaters to communicate directly with the other person and receive feedback immediately, either by voice of text message. The fourth rated was network coverage of the communication device. The boaters in Malaysia compromised with the limitation of network coverage or limited range of their communication device. The fifth rated was the distress signal feature to call the maritime emergency services. Boaters in Malaysia did not think this feature is very important to them because emergency at sea is considered a rare event and they go to the sea in a group of at least two boats for safety reason. Furthermore, they can use their mobile telephone while being within a mobile network's coverage. Price factor was rated sixth. This shows the boaters took the price factor lightly and are not willing to spend considerable money on the communication device. This is the reason why many boaters were using the VHF radio as basic communication device due to its relatively low cost to procure. The seventh and last rated was waterproof factor. Although the communication is an electronic device operating at water environment, waterproof is not considered as very important by the recreational boaters. The communication device can be placed in a sheltered place on the boat such as in the boat's steering wheel cabinet under the boat's canopy. In addition, the waterproof feature can be obtained by using an aftermarket waterproof plastic casing, which is significantly cheaper than buying a waterproof model.

The current communication devices were identified through a literature review. The devices were Garmin inReach Mini, Garmin inReach Explorer, SPOT Gen 4, Medium and High Frequency Radio, VHF Radio, Emergency Positioning Indicating Radio Beacons (EPIRB), Personal Locating Beacon (PLB), and SPOT X Pro and Satellite Phone. A brainstorming session to rate the available maritime communication device was conducted by a group of five experts from MMEA, MDM, nautical

Table 2 Comparison table of best device to be used

Device/ features	Garmin inReach mini	Garmin inReach explorer	SPOT gen 4	MF/HF radio	VHF radio	EPIRB	PLB	SPOT X	Satellite phone iridium	Inmarsat C
Coverage	Pole to pole	Pole to pole	Am, Eu, Au and NZ	Pole to pole	16 km	Pole to pole	Pole to pole	Am, Eu, Au and NZ	Pole to pole	70^0 North to 70^0 South
2-Way Comm	Yes	Yes	No	Yes	Yes	No	No	No	No	Yes
Distress signal	Yes	Yes	Yes	Yes	Yes	Yes	Yes	Yes	Yes	Yes
Portable	Yes	Yes	Yes	Yes	Yes	Yes	Yes	Yes	Yes	Yes
Waterproof	IPX7	IPX7	IPX7	IPX7	IPX7	IPX7	IPX7	IPX7	IPX7	No
Purchase price	USD 350	USD 450	USD 162	USD 4000	USD 160	USD 560	USD 200	USD 200	USD 1050	USD 4,542
Subscription per year	USD 144	USD 144	USD 144	Free	Free	Free	Free	Free	USD 560 (Prepaid)	USD 1085

Note America (AM), Europe (Eu), Australia (Au), and New Zealand (NZ)

Table 3 Rating of communication devices based binary rating system

Device/ features	Garmin inReach mini	Garmin inReach explorer	SPOT gen 4	MF/ HF radio	VHF radio	EPIRB	PLB	SPOT X	Satellite phone iridium	Inmarsat C
Coverage	1	1	−1	1	−1	1	1	−1	1	1
2-Way Comm	1	1	−1	1	1	−1	−1	−1	1	1
Distress signal	1	1	1	1	1	1	1	1	1	1
Portable	1	1	1	−1	0	1	1	1	1	−1
Waterproof	1	1	1	−1	0	1	1	1	1	−1
Purchase price	1	1	1	−1	1	1	1	1	−1	−1
Subscription per year	1	1	1	1	1	1	1	1	−1	−1
Total mark	7	7	3	1	3	5	5	5	3	−1

lecturer, and sailboat owner. Factors used to rate the device were coverage, mode of communication (single or two-way), distress signal, portability, waterproof, purchase price, and subscription plan. Information related to each factor on each device is given in Table 2. Binary rating system was used by the expert group to rate the capability of each device on each factor in Table 3 based on information given in Table 3. The binary numbers –1, 0, and 1 with linguistic value negative, neutral, and positive respectively were used [16]. Based on the rating results, devices were ranked from worst to best as follows: Inmarsat C, MF/HF Radio, VHF Radio, SPOT Gen 4, SPOT X, Satellite Phone Iridium, EPIRB, PLB, Garmin inReach Explorer, and Garmin inReach Mini.

Based on the result obtained, the recommended communication device beyond the mobile telephone network coverage to recreational boaters in Malaysia is the Garmin inReach Mini. The unit cost and subscription fee for the services is within the same range with most mobile phone unit cost and subscription fee in Malaysia. Furthermore, the device uses Iridium satellite that works worldwide and can be linked with mobile a phone for operation. Therefore, the third objective is fulfilled.

5 Conclusion

This study had fulfilled the first objective by identifying three communication devices currently used by the recreational boaters in Malaysia. The devices were marine VHF radio, marine MF/HF radio, and satellite messaging device. The second study objective was fulfilled by identifying the selection factors for the communication device, namely coverage, two-way communication, distress signal, portability, waterproof, and unit cost (price). The third study objective was fulfilled by proposing the Garmin inReach Mini as the most suitable communication device from ten communication devices. The device was selected by using a binary rating system on the six identified factors. Therefore, the satellite messaging device Garmin inReach Mini can improve the safety of recreational boater by providing an effective and low-cost communication system beyond the coverage of mobile telephone communication network.

References

1. Britannica Dictionary, Boating, *Britannica Dictionary* (2022)
2. U.S. Coast Guard, 2020 Recreational Boating Statistics, Washington DC (2021)
3. U.S. Coast Guard, 2019 Recreational Boating Statistics, Washington DC (2020)
4. U.S. Coast Guard, 2018 Recreational Boating Statistics, Washington DC (2019)
5. L.F. Chuah et al., Profiling Malaysian ship registration and seafarers for streamlining future Malaysian shipping governance. Aust. J. Marit. Ocean Aff. **13**(4), 225–261 (2021). https://doi.org/10.1080/18366503.2021.1878981
6. IMO, *SOLAS Consolidated Edition 2009* (IMO, London, 2009)
7. K. Korcz, Maritime radio information system. J. KONES Powertrain Transp. **24**(3), 127–134 (2017). https://doi.org/10.5604/01.3001.0010.3060
8. AMSA, *Australian GMDSS Handbook 2018*, 11th ed. Canberra: AMSA (2018)
9. I. Waugh, *A Mariner's Guide to Marine Communications* (The Nautical Institute, London, 2001)
10. Beachwood Merchandising Corporation, Using VHF Marine Radios (2022). https://quality2wayradios.com/store/using-marine-radios. Accessed 15 Aug 2022
11. P. Chini, G. Giambene, S. Kota, A survey on mobile satellite systems. Int. J. Satell. Commun. Netw. **28**(1), 29–57 (2010). https://doi.org/10.1002/sat.941
12. A.A. Fatah, J.A. Barrett, J. Richard, D. Arcilesi, D.P.S. Scolla, C.H. Lattin, S.D. Fortner, Guide for the selection of communication equipment for emergency first responders. Natl. Inst. Justice **I**(February), 473 (2002)
13. G. Halevi, H. Moed, J. Bar-Ilan, Suitability of google scholar as a source of scientific information and as a source of data for scientific evaluation—review of the literature. J. Informetr. **11**(3), 823–834 (2017). https://doi.org/10.1016/j.joi.2017.06.005
14. A. Joshi, S. Kale, S. Chandel, D. Pal, Likert scale: explored and explained. Br. J. Appl. Sci. Technol. **7**(4), 396–403 (2015). https://doi.org/10.9734/bjast/2015/14975
15. S.F. Zamri, A.F. Ahmad Fuad, M.H. Said, Factor related to operational of marina in west coast peninsular Malaysia. Int. J. e-Navigation Marit. Econ. **15**, 114–122 (2020)
16. I. Pranata, G. Skinner, R. Athauda, A survey on the usability and effectiveness of web-based trust rating systems, *2013 IEEE/ACIS 12th International Conference on Computer and Information Science ICIS 2013—Proceedings*, no. June, pp. 455–460 (2013). https://doi.org/10.1109/ICIS.2013.6607882

Acknowledging the Purpose of Adding Temporary Portable Recycled Hump to Ferries that mitigate the Effects of Roll-Off and Lost-Control Incidents on Passenger Automobiles

Sri Murugan Saga Devan, Noor Apandi Osnin, Bhisham Singh Bhuller, Sofia Irdina Shamiri, and Ain Nabila Mohamad Khalil

1 Introduction

Although speed bumps are intended to slow down traffic in areas with high population density, suburban drivers may find them unpleasant. It is crucial to having barriers in place to prevent speeding. We created the liquid speed hump for this reason, particularly near the ferry ports. This research aims to reduce vehicle damage and avert catastrophic accidents by addressing pertinent issues with conventional concrete speed breakers or other material-based humps [1]. This invention slows down cars, but when it encounters force, the liquid solidifies and forms an obstacle that slows down rapidly moving vehicles. A clever idea that only slows down cars that are traveling too fast is the liquid speed hump [2]. In addition, car access ramps are integrated onto ferry ships. Typically, as seen in Fig. 1, these ramps are constructed toward the stern, or rear, of the ship.

S. M. S. Devan · N. A. Osnin (✉) · B. S. Bhuller · S. I. Shamiri · A. N. M. Khalil
Faculty of Maritime Studies, Universiti Malaysia Terengganu, Terengganu, Malaysia
e-mail: apandi@umt.edu.my

S. M. S. Devan
e-mail: srimurugans04@gmail.com

B. S. Bhuller
e-mail: s64445@ocean.umt.edu.my

S. I. Shamiri
e-mail: s63674@ocean.umt.edu.my

A. N. M. Khalil
e-mail: s63675@ocean.umt.edu.my

A. Fitriadhy et al. (eds.), *Advancing Interdisciplinary Knowledge and Innovation*, SpringerBriefs in Applied Sciences and Technology,
https://doi.org/10.1007/978-3-032-01309-5_2

Fig. 1 Passenger ferries and roll-on roll-off (RORO) ferry services; picture from ministry of transport Malaysia

Problem Statement

Over the preceding 10 years, 915 incidents involving boats hauling cars were reported by the study. These are distributed across Europe's seven hundred ships. Data indicates that this corresponds to about 0.13 accidents per vessel and year. Transported vehicles were reported damaged in 83 of the recorded incidents; a greater number is anticipated. Collisions and cargo that was slick from poor handling were the most frequent causes of mechanical damage to vehicles. A 40-year-old was also driving an RV when it slid down Highway 138 which is located on a steep hill east of the harbor. As the ferry began its ten-minute journey toward Baie-Sainte-Catherine, the automobile crossed the raised loading ramp and struck the back of the vessel. Elec and his passenger were taken to the hospital where his passenger was pronounced deceased. The 40-year-old driver, a native of Levis is still in critical condition, as Fig. 2 illustrates.

Objectives

- To research and enhance a portable, eco-friendly smart hump in a dangerous setting.
- To raise the level of safety by highlighting the importance of drivers driving safely in and out of the vehicle.
- To examine the possibility of a safe driving environment and to reduce dangers in ferry terminals.

Fig. 2 The RV hit the back of the ferry moments after the ship left the Tadoussac terminal on monday. (Submitted by Cindy Desbiens) CBC News Jun 25, 2019

2 Methodology

Material

A non-Newtonian fluid called OBBLECK is considered. If a fluid's viscosity varies in reaction to force or stress, it is said to be non-Newtonian. The most well-known real-world illustration of a non-Newtonian fluid is water dissolved in maize flour. Temperature and pressure alone can be used to characterize Newtonian fluids, such as water. On the other hand, a non-Newtonian fluid's physical behavior is determined by the forces acting on it second to second. Non-Newtonian fluids have several intriguing properties [3].

Non-Newtonian Fluid Classifying Analysis

From a macro perspective, this natural phenomenon may be seen as an inelastic collision, in which the object pressurizes the surrounding liquid to create a solidified plug by drawing the solid–fluid particles together. This nearby liquid then forms a cone-shaped zone around the object when additional force is applied over time. Once the solidification process reaches the bottom of the confining vessel (application of high force), additional forces propagate via a now solid-like area back toward the object, transferring momentum back to the surface and causing it to operate as a hard surface.

(i) The non-linear connection between shear stress and strain rate at a certain location can only be expressed in the following way: $\tau = f\,Y$ where τ and Y are the shear stress and shear rate, respectively. Viscoelastic fluids are those that illustrate partial elastic recovery upon the removal of deforming stress. Such materials possess properties of both fluids and elastic solids and obey Hooke's law of elasticity. For viscous fluids, the constitutive equation can be written as r $Y\ \tau = +\ \pi\ \kappa$ where $\dot{Y}$ and κ are the shear rate and rigidity modulus respectively.

(ii) It is known that the two main categories of time-dependent fluids are thixotropic and rheopectic fluids. These materials have properties of both fluids and elastic solids, and they obey Hooke's law of elasticity.

The rate at which the solidification process starts is determined by the item's speed in the fluid, whose initial value is the object's speed before contact, or let's call it v0. Consequently, if v0 is high enough, the solidification rate becomes apparent and the impacting object experiences a significant magnitude acceleration in the opposite direction of gravity's typical downward path, leading to a somewhat inelastic collision.

Procedure of Fluid Mix Ratio

Mix engineering is the process of selecting suitable ingredients and calculating their ratios to create a non-Newtonian fluid that is as ecologically friendly as feasible while requiring the least amount of workability, strength, and durability. In this experiment, we completely substituted maize starch for concrete. By using the design mix of (3:1.5:1), a non-Newtonian fluid is produced. The strength of the fluid was evaluated with three different substitutions:

The non-Newtonian fluid is prepared using a design mix ratio of

$$\text{Corn Starch} : \text{Water} : \text{Coolant} = 3 : 1.5 : 1$$

The strength of the fluid was tested with the following compositions:

$$\text{Base Mixture} = 85\%\ \text{Corn Starch} + 15\%\ \text{Water}$$

$$\text{Coolant Additions} = 10\%,\ 12.5\%,\ \text{or}15\%\text{of the total mixture}$$

Volume Calculations

Calculate the total capacity of liquid required from the volume of the semi-cylindrical tire tube.

1. Volume of One Capsule

$$Vcapsule = 120\,cm \times 13\,cm \times 0.0102\,cm = 0.0159\,m3$$

2. Total Volume for Two Capsules

$$Vtotal = 2 \times Vcapsule = 2 \times 0.0159\,m3 = 0.0318\,m3$$

Method

Conduit material: When the vehicle wheels over the speed bump, the sealing tubes must be enclosed in a conduit that permits a controlled flow of fluid to enter and exit the chamber and into the reservoir. Synthetic rubber has the following qualities that make it appropriate for this application: 1. Robust elasticity 2. Electrically insulated material 3. Sufficiently resistant to heat, light, and chemicals. 4. High durability against wear and tear. These qualities negates any adverse impacts if pores were to form in the outer shell at some point in the future under extremely hard conditions and exposure.

Outer shell: The material that can be utilized for the outer shell is polyester cloth coated with PVC. These architectural textiles consist of base fabric, adhesive or prime coat, exterior coatings (plasticized PVC), and top coating systems. Polyester coated with PVC has a smooth polyvinyl coating on the back and a fine texture on the front. Since cars are supposed to drive over this shell, the PVC polyester satisfies the critical need of having a high tensile strength. It is waterproof and easily resists ripping and shredding. This vinyl has a covering that includes compounds that are fire-retardant, anti-fungicides, and UV stabilizers.

Base plate: The lightweight bottom plate of the entire design is necessary for the easy relocation of the non-Newtonian fluid speed bumps to the desired place, since they are now movable. The base plate contains fastener holes for bolt and screw attachment. A fiber-reinforced composite plastic panel is the ideal material for the task.

Design

The design of the smart hump is shown in Figs. 3 and 4.

3 Results and Discussion

Liquid Evaporation Test

Tables 1, 2, and 3 and Fig. 5 present the analysis of the water absorption test for 10 days.

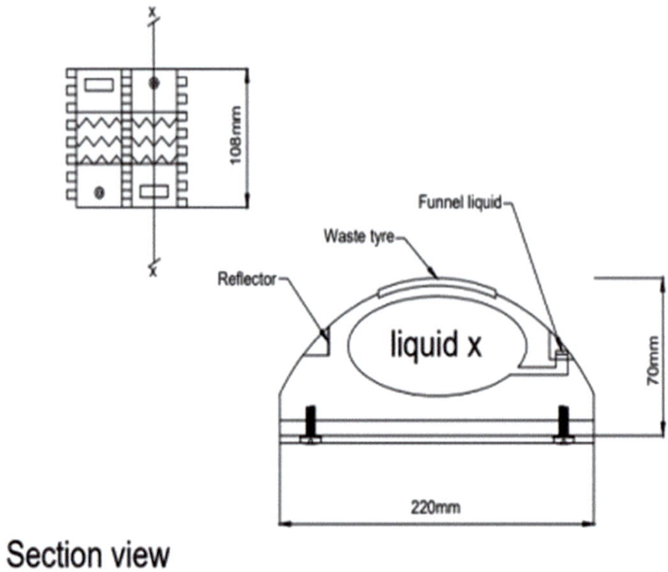

Fig. 3 Section view of the proposed design

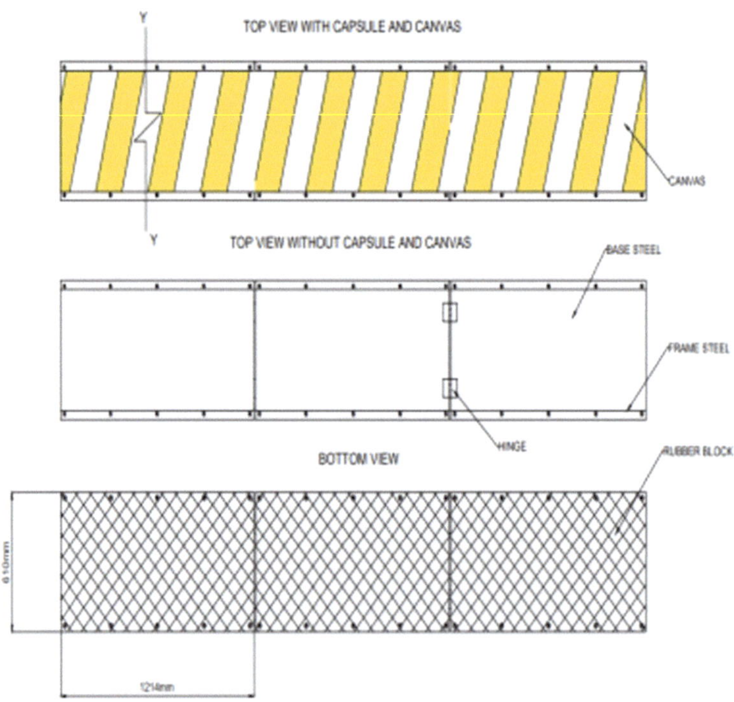

Fig. 4 Section view of the layout of top view and bottom view of the portable smart hump

Table 1 Percentage and weight for each portion of the substance

Material	Density (kg/m^3)	Weight of each portion used (kg)			The percentage of cement and silica fume
Corn starch	625	3	3	3	(85/15) (80/20) (50/50)
Water	1000	1.5	2.5	3	
Coolant	1.1132	1	0.75	1.25	
The percentage for corn starch and coolant %		(60/40)	(70/30)	(80/20)	

Table 2 Actual volume-based capacity

Material	Density (kg/m^3)	Weight of each portion used (kg/litre)
Corn starch	625	8.69
Water	1000	4.34
Coolant	1.1132	2.89

Table 3 Calculation method for each portion of the substance

Material	Method of calculation
Corn starch and water	Percentage of material ÷ overall ratio (3) × density × volume
Coolant and mixture	Percentage of material x ratio (2) ÷ overall ratio (3) × density × volume

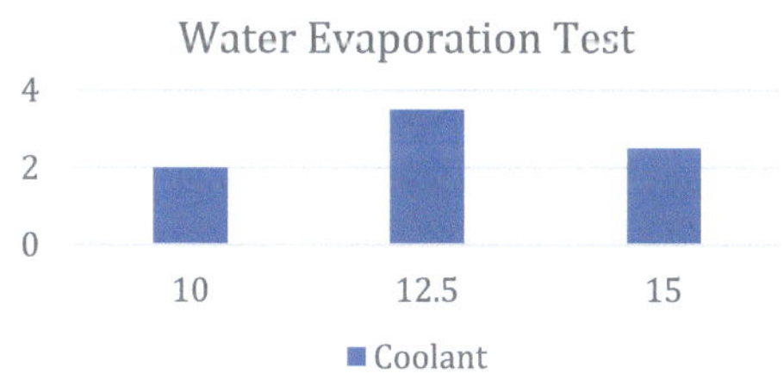

Fig. 5 Water evaporation reactions for different types of coolants

Discussion

Numerous studies that looked at the effectiveness of smart humps found that they would be very helpful in minimizing roll-off incidents involving cars that lost control of ferries. This transportable recycling smart hump can decrease accidents and increase public awareness. The negative effects are many and include things like speed limit enforcement, damage to automobiles, and obstructing emergency vehicles [4]. Vehicle loading and unloading are governed by clear regulations, and there are established safety protocols for using the boats. Keep passenger cars up to date with safety regulations. Eventually, this ingenious hump was created, either to enhance the original concept or to modify it for a little different use. To enhance vehicular traffic in a secure setting, vessels, such as ferries and port areas, should have a mobile recycling smart hump. Offering flexible and adaptive strategic planning for their assets—especially multifunctional ferries—is the aim of this research. To enhance vehicle traffic in a safe environment, smart hump installations are required for boats such as ferries and port facilities. The goal of this study is to develop strategic planning that is adaptive and flexible for their assets, particularly multifunctional ferries [5].

4 Conclusion

Smart humps are still in use today, but their initial purpose has been modified to only slow down individuals who require it, thanks to ingenious material technology. By ensuring ecologically responsible operations, the liquid smart hump will promote drivers' awareness of safety concerns. This is a perfect example of how new technologies may improve workers' driving skills while lowering risk and improving environmental safety. By lowering vehicle speeds, safeguarding public safety, lowering fatal accidents, and lowering noise pollution, the modern Oobleck non-Newtonian fluid speed bump achieves its intended purpose of lowering vehicle speeds. It is without a doubt a safer and greener option than the conventional speed breaker. Mobility and emergency detection technologies are added to the design, making it even more futuristic and effective in ensuring public safety. By using ecologically friendly composite materials, the project's sustainability is increased. This study paper lays the groundwork for the plan's creation, but before this concept can be implemented, other issues must be addressed.

Acknowledgements The authors would like to thank the Department of Civil Engineering, Faculty of Maritime Studies, University Malaysia Terengganu, and everyone who helped with this study.

References

1. T.A. Abdel-Wahed, I.H. Hashim, Effect of speed hump characteristics on pavement condition. J. Traffic Transp. Eng. (Engl. Ed.) **4**(1), 103–110 (2017)
2. M. Albany, M. Sadek, Analysis of speed hump effect on signalized inter section performance. Port-Said Eng. Res. J. **25**(1), 21–31 (2021)
3. Marine Accident Investigation Branch. Cargo shift and damage to vehicles on board ro-ro passenger ferry European Causeway (2020). https://assets.publishing.service.gov.uk/media/5e1dc77f40f0b65dc191878a/2020-03-EuropeanCauseway.pdf. Accessed 15 Jul 2023
4. Oobleck.: A non-Newtonian fluid revolutionizing speed bumps (2021). https://eduindex.org/2021/06/03/oobleck-revolutionizing-speed-bumps/. Accessed 25 Jul 2023
5. Real Life Incident: Ro-Ro Vessel Rolls Heavily and Tips Trucks (2020). https://www.marineinsight.com/shipping-news/real-life-incident-ro-ro-vessel-rolls-heavily-and-tips-trucks/. Accessed 25 Jul 2023

A Credit Score in Pay Later Service by Using Machine Learning Optimization: A Bibliometric Analysis

Yosy Arisandy, Yosza Dasri, and Shahrul Nizam Salahudin

1 Introduction

Individual wellbeing promotes good society values, and in turn good society values are results from many inter-related factors [1]. Indeed, one of those factors is debt management among individuals. Considering recent trends in personal debt accumulation, there has been a rise in the utilization of a type of short-term financing referred to as buy now pay later (BNPL). This financial arrangement allows consumers to make purchases online and defer payment to a later date, often without incurring any interest charges.

Credit scoring models are utilized in order to forecast the likelihood of a client defaulting on credit at a specific point in time. These models rely on the personal and financial information of the consumer, which can impact their capacity to repay the debt [2, 3]. This has become a subject of increasing apprehension [4]. The dynamic nature of the banking industry poses challenges to many sectors within it, as it hampers the efficacy of data investigation processes. These processes heavily rely on human procedures and are often hindered by the continuous changes occurring in the data produced and stored in computer systems [5].

Y. Arisandy (✉) · S. N. Salahudin
Faculty of Technology Management and Business, Universiti Tun Husein Onn Malaysia, Johor, Malaysia
e-mail: gp200002@student.uthm.edu.my

S. N. Salahudin
e-mail: shahrulns@uthm.edu.my

Y. Dasri
Faculty of Electronics & Computer Technology & Engineering, Universiti Teknikal Malaysia Melaka, Durian Tunggal 76100, Melaka, Malaysia
e-mail: yosza@utem.edu.my

A. Fitriadhy et al. (eds.), *Advancing Interdisciplinary Knowledge and Innovation*, SpringerBriefs in Applied Sciences and Technology,
https://doi.org/10.1007/978-3-032-01309-5_3

A large number of peer-to-peer lending platforms allow individuals the opportunity to obtain a loan by either the provision of collateral or the assessment of a "credit score." This score is determined by considering various factors, including the user's credit history [6], financial circumstances [7], and pertinent information pertaining to loans and debts [8]. Credit scoring uses the timely and appropriate payment of bills by consumers to determine how they pay off their debts [9].

The core objectives of this study are to determine the quantity of publications and the dissemination of research findings pertaining to credit scores, deferred payment, and payment behavior within the designated study time frame. Additionally, this research seeks to present a bibliometric and mapping analysis of credit score measurements identified through an extensive review of relevant literature. In order to ascertain the extent of the gap and uniqueness, bibliometric analyses are employed. Bibliometric analyses can be employed to assess the progression of a trend or the publication output by examining the quantity of publications within specified subject fields [10].

2 Methodology

This research utilizes bibliometric analysis using data sources from Scopus. Scopus is widely recognized as a preeminent and influential repository of high-quality scholarly literature, particularly in comparison to other index databases [10, 11]. The research database was utilized on the website www.scopus.com, using the specified criteria of examining the title, abstract, and keywords "Credit Score," "Pay Later," and "Payment Behavior" throughout the time frame of 2017 to 2022. There are various limitations that exist within the realm of research. These limitations encompass the fields of computer science, business, management, and accounting, as well as economics, econometrics, finance, engineering, social sciences, decision science, and environmental science. The types of documents that are commonly encountered within these fields include articles, conference papers, and book chapters. Furthermore, it is important to note that the primary language used in these academic pursuits is English. A total of 317 results were obtained. This study focuses on the utilization of VOSviewer and RStudio for the processing of research data.

3 Results and Discussion

The primary information, in the form of data, is presented in the subsequent section. Table 1, which was generated using RStudio, presents a lot of significant data. The document's body contains two distinct terms: Keywords Plus and Author's Keywords (DE). Keywords Plus refers to the frequency distribution of keywords associated with a specific document inside the databases of Clarivate Analytics Web of Science and Scopus. The distribution of the frequency of the writers' keywords is referred to as

Author's Keywords [12]. The ranking of document types for this particular dataset is as follows: articles rank first with 65% representation and a total of 207 documents, conference papers rank second with 31% representation and a total of 97 documents, and book chapters rank fourth with 4% representation and a total of 13 documents.

According to Fig. 1 Annual Scientific Production, there is a growing trend in the exploration and popularity of themes such as "credit score," "pay later," and "customer behavior" over the years. Nevertheless, the current level of publication on this topic in 2022 remains rather low due to the limited availability of data collected just till August. However, it is important to note that this level is still higher compared to that of 2017. The most prolific sources of research on this topic are the ACM International Conference Proceeding Series and Advances in Intelligent Systems and Computing, both of which have 10 publications each.

The present study examines the works of ten prominent authors who have extensively written on the subject matter. In the process of data processing using RStudio, Zhang Y produced a total of six documents on the Scopus page during the specified search period. This was followed by Li H, Li Y, and Liu Y, who each generated four

Table 1 Main information about data

Description	Results
Main information about data	
Timespan	2017:2022
Sources (Journals, Books, etc.)	233
Documents	317
Annual growth rate %	7.47
Document average age	2.26
Average citations per doc	6.107
References	12,426
Document contents	
Keywords plus (ID)	1327
Author's keywords (DE)	1083
Authors	
Authors	865
Authors of single-authored docs	42
Authors collaboration	
Single-authored docs	44
Co-Authors per Doc	2.99
International co-authorships %	18.61
Document types	
Article	207
Book chapter	13
Conference paper	97

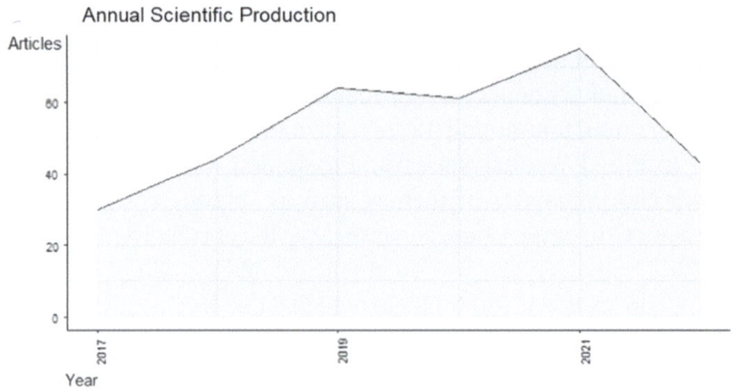

Fig. 1 Annual scientific production

documents, following with Chen K, Chen Y, Giudici P, Kumar N, Kiviat B, and Liu H, each possessing a collection of three documents.

The United States is recognized as a nation that exhibits a substantial volume of publications pertaining to the subject matter, specifically 209 documents, as depicted in Fig. 2 Country Scientific Production statistics. Conversely, China holds the distinction of achieving the greatest count in generating scholarly works inside this particular study domain, with a total of 192 documents. India ranked second with a total of 121 documents. The quantity of publications in Indonesia and England is equivalent, with a total of 30 documents each. This finding suggests that Indonesia holds the largest number of publications in this discipline among Southeast Asian nations. In Europe, Italia ranks among the leading countries in scientific production, having contributed 25 documents, while Germany follows closely with 23 documents. South Korea ranked eighth, having produced 23 documents, while Brazil and Turkey followed closely with 19 and 17 documents respectively.

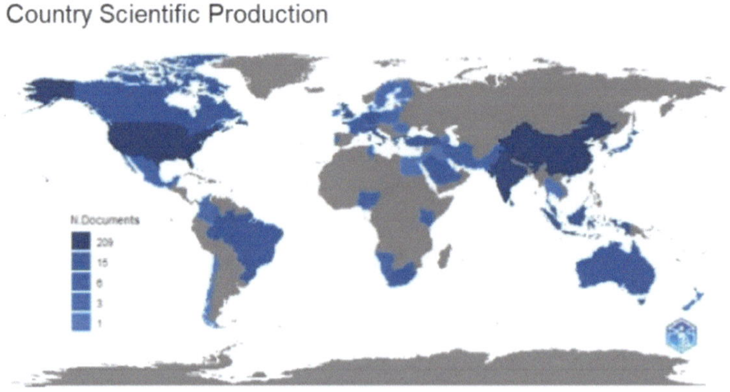

Fig. 2 Country science production

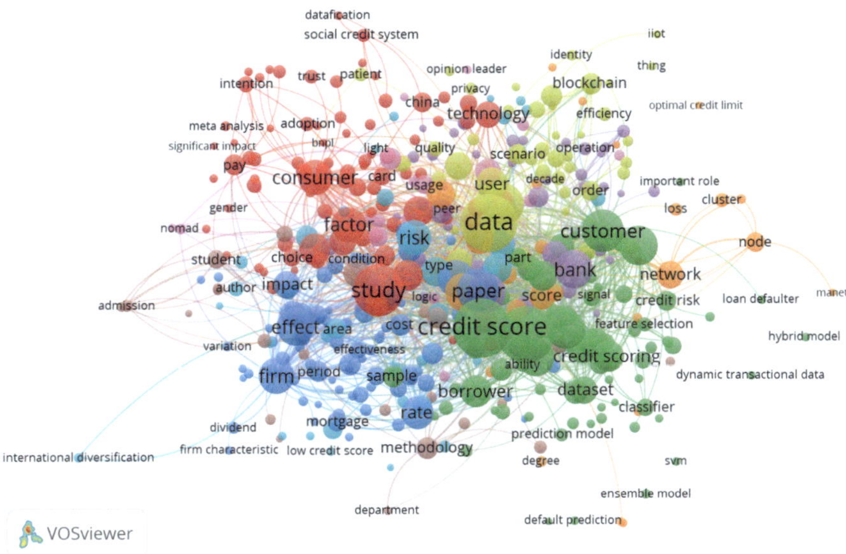

Fig. 3 Network visualization using Vosviewer

Based on the analysis conducted using VOSviewer or RStudio, it has been seen that the term “credit score” exhibits a high frequency of occurrence, as depicted in Figs. 3 and 4. In the context of data processing using VOSviewer, it is seen that there is a lack of information regarding the frequency and count of words. Conversely, the utilization of a network plot in RStudio enables the provision of both the amount and percentage of data occurrences. Based on the gap analysis derived from network analysis, it is apparent that there has been a substantial amount of study undertaken on credit ratings. However, a notable void exists in terms of establishing a network that links credit scores to pay later services. Furthermore, there is a lack of connectivity between the pay later service, machine learning, and feature optimization, so that academics can relate the keywords credit score, pay later service, and machine learning in the future study.

In relation to the subject of trend analysis, it is observed that research pertaining to the keyword “Credit scoring” has been predominantly conducted, with a frequency of 43. Additionally, the term “credit score” has been cited 29 times in the years 2019, 2020, and 2021. The subjects of machine learning and scoring systems continue to be widely discussed and studied up until the year 2022. The terms “Payment behavior” and “Future selection” alone emerged as significant keywords throughout the period from 2019 to 2020. Hence, the investigation pertaining to credit ratings is firmly established and is a prominent subject of interest within the scientific community.

The thematic map depicted according to Fig. 5 comprises four quadrants, commonly referred to as strategic diagrams. The quadrant located in the upper right region is sometimes referred to as the motor theme or driving issue. The lower left quadrant is designated as “declined” or “emerging,” encompassing research that is

Fig. 4 Word cloud using RStudio

experiencing a decline or perhaps showing signs of growth. As one ascends, the concentration of journals or the quantity of journals pertaining to this subject matter increases, with a corresponding shift toward the right in their display. Centrality, usually referred to as the expanding influence of the subject on the field, is a well-established concept. The top left quadrant of the diagram suggests the presence of multiple publications, yet with distinct divisions resulting from a limited number of comprehensive investigations. The presence of a substantial influence on other articles is indicated by the bottom right quadrant, suggesting that the issue holds great importance. The theme map clearly indicates that the discourse surrounding credit ratings and machine learning continues to be a subject of significant scholarly attention.

Table 1 reveals that machine learning is employed in 37 documents, constituting 27% of the credit score computation process. A total of six scholarly articles employs a combination of machine learning techniques and alternative approaches for measuring credit scores, such as data mining, neural networks, artificial intelligence, regression analysis, and other measurement tools. Furthermore, machine learning is employed in 30 additional documents for this purpose. There are several factors contributing to the widespread adoption of machine learning. Firstly, machine learning techniques are increasingly being utilized for classification-based problems [13]. Additionally, machine learning offers more comprehensive solutions and has the potential to enhance the accuracy and structure of credit scoring systems [14]. Moreover, machine learning is extensively employed to address both supervised and unsupervised problems [15]. The integration of machine learning can be conceptualized as a strategy to enhance the precision and accuracy of the system [14].

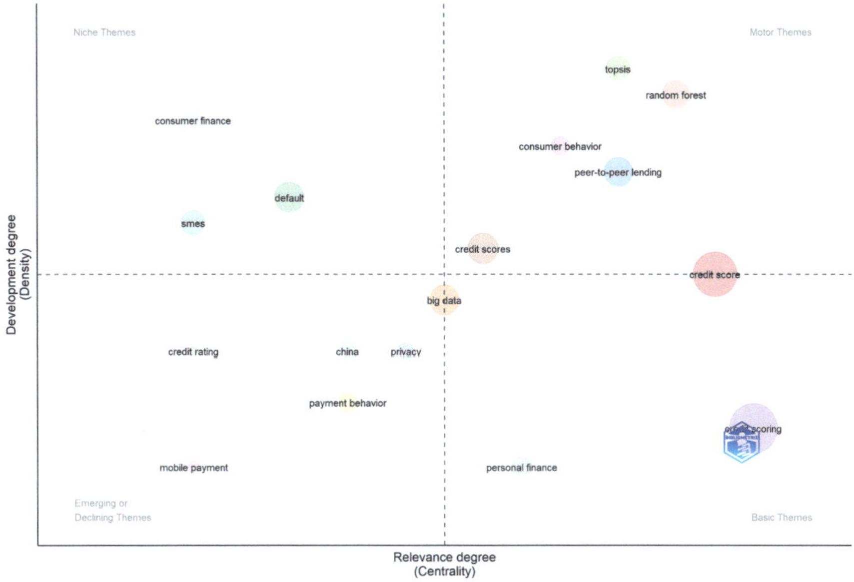

Fig. 5 Thematic map using RStudio

Table 3 shows the classifications of database sources utilized for data processing, which are generated from the machine learning research described in Table 2.

Most research studies on credit scores utilizing machine learning techniques primarily concentrate on binary classification tasks, specifically distinguishing between paying and non-paying individuals [97], as well as identifying problematic creditors and excellent creditors [98]. However, it is also possible to categorize it as multi-class classification or using statistical methods such as chi-square analysis in conjunction with another credit score technique.

Machine learning is currently regarded as the most efficient approach for processing data. The utilization of mobile phone usage data has the potential to enhance credit scoring, even when working with limited datasets. Our study provides evidence that the decision tree method [97] is the most effective model for credit scoring in relation to mobile data consumption. The field of machine learning [16] has promise for future investigations. The utilization of machine learning approaches plays a crucial role in the detection of credit risk [98], particularly in predicting those who are likely to default payment. The integration of batch and stream learning techniques can yield improved credit scoring systems, while also facilitating the discovery of pertinent gaps that necessitate additional investigation. In developing nations, where consumer behavior evolves over time, this amalgamation has the potential to yield positive outcomes and warrants additional investigation and implementation [17]. According to previous research [18], it has been demonstrated that machine learning (ML) classification algorithms possess the capability to effectively and accurately classify the credit risk of borrowers.

Table 2 Credit score measurement

NO	Credit score measurement	Source	
1	Machine learning	[13, 16– 49]	
2	Artificial intelligence and machine learning	[50]	
3	Data mining and machine learning	[51]	
4	007fenqi's credit score system	[52]	
5	Data mining	[53, 54]	
6	Multiple regression techniques	[55]	
7	Analytical hierarchy process and fuzzy	[56]	
8	Artificial neural networks	[57– 60]	
9	Artificial intelligence	[61– 63]	
10	Binary bat algorithm; heuristic algorithm and radial basis function neural network (RBFN)	[64]	
11	Blockchain	[65]	
12	Cross-sectional regression model	[66]	
13	C-SCORE	[67]	
14	Descriptive statistics and regression analysis	[68]	
15	Neural network used genetic algorithm-LSTM technique	[69]	
16	Ethereum test network	[70]	
17	FICO	[71]	
18	FinRep Analyser	[72]	
19	Fuzzy	[73– 76]	
20	L-M optimization algorithm; neural network	[77]	
21	Linear regression model	[78]	
22	Logistic regression	[79– 86]	
23	Neural network used semi-supervised learning (SSL)	[87]	
24	Myers-Briggs type indicator (MBTI) personality types	[88]	

(continued)

Table 2 (continued)

NO	Credit score measurement	Source	
25	Nature-inspired algorithms; evolutionary algorithms (EAs)	[89]	
26	SeQuiScore credit indicator	[90]	
27	Smartphone usage using multi-cluster feature selection (MCFS)	[91]	
28	Spectral clustering and cost-sensitive deep neural network	[92]	
29	Spreadsheet modeling and database querying	[93]	
30	The Kruskal–Wallis non-parametric statistic	[57]	
31	The NCB score, based on credit behavior information by the national credit bureau of Thailand (NCB)	[94]	
32	The statistical tools chi-square analysis and correspondence analysis	[95]	
33	Voice analysis; the voice waveform and pitch analyses	[96]	102
34	Undefined		35

Table 3 Machine learning database source

No	Database source in machine learning processing data to defined credit score
1	Machine learning library [16]
2	Survey; background, smartphone usage and trustworthiness [91]
3	Online social networks [36]
4	Financial data company [34]
5	Social lending platform (Lending Club) dataset [39]
6	Combine primary (survey) and secondary data [7]

Despite its disadvantages, pure and fundamental machine learning methods currently have several constraints when it comes to handling time-varying input data. These limitations restrict their capacity to effectively include future and unseen classes, which in turn hampers their ability to generate generalization and knowledge from the available data [99].

4 Conclusion

With the expanding scope of financial offerings and the diversification of financial service providers, traditional banks are no longer the exclusive purveyors of financial products. Consequently, novel and inventive financial instruments, such as unsecured loans like pay later options, have emerged in the market. Consequently, there has been a notable increase in the volume of study in this field over the years, with a total of 317 articles recorded. Machine learning is a prevalent methodology employed for the assessment of credit scores. The ACM International Conference Proceedings Series and Advances in Intelligent Systems and Computing are widely recognized as significant sources of research in this field. Notably, the United States emerges as the primary contributor to these publications. Zhang Y is an exceptionally productive author. The analysis conducted using VOSviewer and RStudio revealed that previous studies have focused on applying machine learning techniques to calculate credit scores in various domains such as credit card analysis, P2P lending, and loan banking systems. However, there is a notable gap in the literature as no specific research has been conducted on the application of machine learning in the context of pay later services. Furthermore, the categorization performed based on credit ratings primarily consists of two classifications: positive and negative credit performances. This study proposes that future investigations on credit ratings utilizing machine learning techniques could be broadened to encompass pay later services, incorporating innovative supplementary metrics. These metrics could consist of five distinct levels, namely terrible, acceptable, good, very good, and exceptional. Consequently, lenders should have the ability to assess borrowers with nonperforming loans in order to issue a cautionary notice when the borrower's credit score is nearing a diminished threshold and to find prospective borrowers.

Acknowledgements The research conducted in this study received financial support from Universiti Tun Hussein Onn Malaysia through the TIER1 grant, with reference number H777.

References

1. S.N. Salahudin, S.S. Baharuddin, M.N.R. Alwi, Conceptual framework of determinants well-being of orang asli in Malaysia **23**(9) (2017). https://doi.org/10.1166/asl.2017.9947
2. K.P. Wong, On Determ. Bank Interes. Margins Credit. Interes. Rat. Risks **21**(2), 251–271 (1997). https://doi.org/10.1016/S0378-4266(96)00037-4
3. A. Mendonça, *The Unconventional Monetary Policy of the ECB and the International Economic and Financial Crisis: Effectiveness Versus Exhaustion*, pp. 223–247 (2017). https://doi.org/10.1007/978-3-319-45710-9_14
4. A.I. Baranes, C. Coward Bucher, Classification systems and the Veblenian dichotomy: an interdisciplinary approach to social stratification **56**(2), 570–577 (2022). https://doi.org/10.1080/00213624.2022.2065855
5. T. Harris, Credit. Scoring Using Clust. Support. Vector Mach. **42**(2), 741–750 (2015)
6. J. Hartmann, O. Hasan, A social-capital based approach to blockchain-enabled peer-to-peer lending, in *2021 3rd International Conference on Blockchain Computing and Applications, BCCA 2021*, pp. 105–110 (2021). https://doi.org/10.1109/BCCA53669.2021.9656964
7. A. Senthil Kumar, S. Ramesh, S. Rahul, A technology on credit score system-assessing public perception in Bengaluru city **8**(12), 1929–1934 (2019). https://doi.org/10.35940/ijitee.L2890.1081219
8. Y. Ji, K. Chalkias, Generalized proof of liabilities, in *Proceedings of the ACM Conference on Computer and Communications Security*, pp. 3465–3486 (2021). https://doi.org/10.1145/3460120.3484802
9. P. Credit Bureu, How Does Payment Behavior Affect Credit Score? (2022). https://www.idscore.id/education/detail/bagaimana-perilaku-pembayaran-berpengaruh-pada-credit-score. Accessed 10 Oct 2022
10. W.A.N. Norhidayah, H. Suziana, Mapping major trends in global research in marine and coastal tourism : a bibliometric analysis **17**(8), 200–217 (2022). https://doi.org/10.46754/jssm.2022.08.012

11. A. Ahmi, M.H. Mohd Nasir, Examining the trend of the research on extensible business reporting language (xbrl): a bibliometric review **5**(2) (2019)
12. M. Aria, C. Cuccurullo, Bibliometrix: an R-tool for comprehensive science mapping analysis **11**(4), 959–975 (2017). https://doi.org/10.1016/j.joi.2017.08.007
13. D. Swain, R. Agrawal, A. Chandak, V. Lapshetwar, N. Chandak, A. Vaswani, *Credit Score Prediction Using Machine Learning*, 1311 AISC (2021). https://doi.org/10.1007/978-981-33-4859-2_7
14. M. Rudra Kumar, V.K. Gunjan, *Peer Level Credit Rating: An Extended Plugin for Credit Scoring Framework*. vol. 828 (2022). https://doi.org/10.1007/978-981-16-7985-8_128
15. M. Rabbani, A. Aghsami, S. Farahmand, S. Keyhanian, Risk Revenue Lessor's Dyn. Jt. Pricing Inven. Plan. Adjust. Costs Differ. Inflat. **11**(1), 1–35 (2018). https://doi.org/10.1504/IJPM.2018.088613
16. P. Singh, Comparative study of individual and ensemble methods of classification for credit scoring, in *Proceedings of the International Conference on Inventive Computing and Informatics, ICICI 2017*, pp. 968–972 (2018). https://doi.org/10.1109/ICICI.2017.8365282
17. J. P. Barddal, F. Enembreck, L. Loezer, R. Lanzuolo, Combining slow and fast learning for improved credit scoring, in *Conference Proceedings—IEEE International Conference on Systems, Man and Cybernetics*, pp. 1149–1154 (2020). https://doi.org/10.1109/SMC42975.2020.9283453
18. L. Guntay, E. Bozan, U. Tigrak, T. Durdu, G.E. Ozkahya, An explainable credit scoring framework: a use case of addressing challenges in applied machine learning, in *2022 IEEE Technology and Engineering Management Conference: Societal Challenges: Technology, Transitions and Resilience Virtual Conference, TEMSCON EUROPE 2022*, pp. 222–227 (2022). https://doi.org/10.1109/TEMSCONEUROPE54743.2022.9802029
19. O.J. Okesola, K.O. Okokpujie, A.A. Adewale, S.N. John, O. Omoruyi, An improved bank credit scoring model: a Naïve Bayesian approach, in *Proceedings—2017 International Conference on Computational Science and Computational Intelligence, CSCI 2017*, pp. 228–233 (2018). https://doi.org/10.1109/CSCI.2017.36
20. M. Herasymovych, K. Märka, O. Lukason, Using reinforcement learning to optimize the acceptance threshold of a credit scoring model **84** (2019). https://doi.org/10.1016/j.asoc.2019.105697
21. Y. Li, Y. Sun, N. Contractor, Graph mining assisted semi-supervised learning for fraudulent cash-out detection, in *Proceedings of the 2017 IEEE/ACM International Conference on Advances in Social Networks Analysis and Mining, ASONAM 2017*, pp. 546–553 (2017). https://doi.org/10.1145/3110025.3110099
22. Y. Xia, X. Yang, Y. Zhang, A rejection inference technique based on contrastive pessimistic likelihood estimation for P2P lending **30**, 111–124 (2018). https://doi.org/10.1016/j.elerap.2018.05.011
23. H. Xia, Y. Wang, J. Gauthier, J.Z. Zhang, Knowledge graph of mobile payment platforms based on deep learning: risk analysis and policy implications **208** (2022). https://doi.org/10.1016/j.eswa.2022.118143
24. P. DIaconescu, V.-E. Neagoe, Credit scoring using deep learning driven by optimization algorithms, in *Proceedings of the 12th International Conference on Electronics, Computers and Artificial Intelligence, ECAI 2020* (2020). https://doi.org/10.1109/ECAI50035.2020.9223139
25. V. Veeramanikandan, M. Jeyakarthic, Parameter-Tuned Deep. Learn. Model. Credit. Risk Assess. Scoring Appl. **14**(9), 2958–2968 (2021). https://doi.org/10.2174/2666255813999200819164013
26. A. Pandey, S. Shukla, K.K. Mohbey, Comp. Anal. Deep. Learn. Approach Var. Cl. Tech. Credit. Score Comput. **14**(9), 2785–2799 (2021). https://doi.org/10.2174/2666255813999200721004720
27. D. Babaev, A. Tuzhilin, M. Savchenko, D. Umerenkov, E.T.-Rnn: applying deep learning to credit loan applications, in *Proceedings of the ACM SIGKDD International Conference on Knowledge Discovery and Data Mining*, pp. 2183–2190 (2019). https://doi.org/10.1145/3292500.3330693

28. K. Yang, H. Yuan, R.Y.K. Lau, PsyCredit: an interpretable deep learning-based credit assessment approach facilitated by psychometric natural language processing **198** (2022). https://doi.org/10.1016/j.eswa.2022.116847
29. S.B. Patel, P. Bhattacharya, S. Tanwar, N. Kumar, KiRTi: a blockchain-based credit recommender system for financial institutions **8**(2), 1044–1054 (2021). https://doi.org/10.1109/TNSE.2020.3005678
30. S.K. Trivedi, A study on credit scoring modeling with different feature selection and machine learning approaches **63** (2020). https://doi.org/10.1016/j.techsoc.2020.101413
31. A. Shema, Effective credit scoring using limited mobile phone data. ACM Int. Conf. Proceeding Ser. (2019). https://doi.org/10.1145/3287098.3287116
32. M. Pincovsky, A. Falcao, W.N. Nunes, A.P. Furtado, R.C.L.V. Cunha, Machine learning applied to credit analysis: a systematic literature review, in *Iberian Conference on Information Systems and Technologies, CISTI* (2021). https://doi.org/10.23919/CISTI52073.2021.9476350
33. Z. Wen et al., Customer churn warning with machine learning, vol. 891. Springer Verlag, Bank of China, Beijing, 100091, China, pp. 343–350 (2019). https://doi.org/10.1007/978-3-030-03766-6_39
34. A. M. Christidi, C. Christidis, T. Manolatos, I. Hatzilygeroudis, E. Voyiatzaki, Modelling prediction of enterprises payment behavior for applying machine learning methods, University of Patras, Department of Computer Engineering Informatics, Patras, Greece: Institute of Electrical and Electronics Engineers Inc. (2019). https://doi.org/10.1109/IISA.2019.8900680
35. Y. Lu, Z. Sheng, K. Zhang, Q. Duan, Predicting P2P lenders' decisions: the prospect theory approach **17**(1), 1–12 (2022). https://doi.org/10.1504/IJEB.2022.119935
36. M. Deeptha, G.S. Arpitha, S.D. Anjali, D.B. Srinivas, *UCREDIT—Credit Scoring Using Social Media Platforms* **790** (2022). https://doi.org/10.1007/978-981-16-1342-5_31
37. H. Gao, H. Liu, H. Ma, C. Ye, M. Zhan, Netw.-Aware Credit. Scoring Syst. Telecom Subscr. Using Mach. Learn. Netw. Anal. **34**(5), 1010–1030 (2022). https://doi.org/10.1108/APJML-12-2020-0872
38. V.B. Djeundje, J. Crook, R. Calabrese, M. Hamid, Enhancing credit scoring with alternative data **163** (2021). https://doi.org/10.1016/j.eswa.2020.113766
39. V. Moscato, A. Picariello, G. Sperlí, A benchmark of machine learning approaches for credit score prediction **165** (2021). https://doi.org/10.1016/j.eswa.2020.113986
40. A. Tunç, Feature selection in credibility study for finance sector, in *Procedia Computer Science*, pp. 254–259 (2019). https://doi.org/10.1016/j.procs.2019.09.049
41. S. Kokate, M.S.R. Chetty, Credit risk assessment of loan defaulters in commercial banks using voting classifier ensemble learner machine learning model **11**(5), 565–572 (2021). https://doi.org/10.18280/IJSSE.110508
42. M.P. Kim, A. Ghorbani, J. Zou, Multiaccuracy: black-box post-processing for fairness in classification, in *AIES 2019—Proceedings of the 2019 AAAI/ACM Conference on AI, Ethics, and Society*, pp. 247–254 (2019). https://doi.org/10.1145/3306618.3314287
43. G. Teles, J.J.P.C. Rodrigues, K. Saleem, S.A. Kozlov, Cl. Methods Appl. Credit. Scoring Collat. **14**(3), 4557–4566 (2020). https://doi.org/10.1109/JSYST.2019.2937552
44. X. Shi, K. Chen, J. Zhang, Research on consumer credit portrait intelligent scoring model, in *ACM International Conference Proceeding Series*, pp. 46–51 (2020). https://doi.org/10.1145/3399871.3399901
45. H. Li et al., A Nov. Method Credit. Scoring Based Featur. Transform. Ensemble Model. **7**, 1–18 (2021). https://doi.org/10.7717/PEERJ-CS.579
46. J. P. Barddal, L. Loezer, F. Enembreck, R. Lanzuolo, Lessons learned from data stream classification applied to credit scoring **162** (2020). https://doi.org/10.1016/j.eswa.2020.113899
47. H. Wang, W. Chen, F. Da, Zhima credit score in default prediction for personal loans, in *Procedia Computer Science*, pp.1478–1482 (2021). https://doi.org/10.1016/j.procs.2022.01.188

48. D. Cheng, Z. Niu, Y. Zhang, Contagious chain risk rating for networked-guarantee loans, in *Proceedings of the ACM SIGKDD International Conference on Knowledge Discovery and Data Mining*, pp. 2715–2723 (2020). https://doi.org/10.1145/3394486.3403322
49. S. Barua, D. Gavandi, P. Sangle, L. Shinde, J. Ramteke, Swindle: predicting the probability of loan defaults using CatBoost algorithm, in *Proceedings—5th International Conference on Computing Methodologies and Communication, ICCMC 2021*, pp.1710–1715 (2021). https://doi.org/10.1109/ICCMC51019.2021.9418277
50. R. Njuguna, K. Sowon, Poster: A scoping review of alternative credit scoring literature, in *Proceedings of 2021 4th ACM SIGCAS Conference on Computing and Sustainable Societies, COMPASS 2021*, pp. 437–444 (2021). https://doi.org/10.1145/3460112.3471972
51. P.K. Giri, S.S. De, S. Dehuri, S.-B. Cho, Biogeography based optimization for mining rules to assess credit risk. Intell. Syst. Account. Finan. Manag. **28**(1), 35–51 (2021). https://doi.org/10.1002/isaf.1486
52. C. Leong, B. Tan, X. Xiao, F.T.C. Tan, Y. Sun, Nurturing a FinTech ecosystem: The case of a youth microloan startup in China. Int. J. Inf. Manage. **37**(2), 92–97 (2017). https://doi.org/10.1016/j.ijinfomgt.2016.11.006
53. S. Carta, A. Ferreira, D. Reforgiato Recupero, M. Saia, R. Saia, A combined entropy-based approach for a proactive credit scoring. Eng. Appl. Artif. Intell. **87** (2020). https://doi.org/10.1016/j.engappai.2019.103292
54. R.H. Mahjoub, A. Afsar, A hybrid model for customer credit scoring in stock brokerages using data mining approach. Int. J. Bus. Inf. Syst. **31**(2), 195–214 (2019). https://doi.org/10.1504/IJBIS.2019.100279
55. D. Kim, Empirical evidence of faulty credit scoring and business failure in P2P lending. Glo. Bus. Finan. Rev. **26**(2), 67–82 (2021). https://doi.org/10.17549/gbfr.2021.26.2.67
56. J. Yu, J. Yao, Y. Chen, Credit scoring with AHP and fuzzy comprehensive evaluation based on behavioural data from weibo platform. Tehnicki Vjesnik **26**(2), 462–470 (2019). https://doi.org/10.17559/TV-20181217180231
57. A. Ashofteh, J.M. Bravo, A non-parametric-based computationally efficient approach for credit scoring. Atas Da Conferencia Da Associacao Portuguesa de Sistemas de Informacao (2019)
58. Y. Li, T. Wang, M. Kang, Y. Zhang, Y. Chen, Understanding users' intentions towards the internet consumer credit platform: Evidence from China (2017). https://www.scopus.com/inward/record.uri?eid=2-s2.0-85091084521&partnerID=40&md5=d0c399df220485811d988b0a0b048e3b
59. J. Lohokare, R. Dani, S. Sontakke, Automated data collection for credit score calculation based on financial transactions and social media, in *2017 International Conference on Emerging Trends and Innovation in ICT, ICEI 2017*, pp. 134–138 (2017). https://doi.org/10.1109/ETIICT.2017.7977024
60. L.K. Maina, B.S. Kasamani, A centralized credit scoring prototype for microlending institutions using neural networks, in *Lecture Notes in Networks and Systems*, vol. 393 (2022). https://doi.org/10.1007/978-3-030-94191-8_11
61. T. Handhika, A. Fahrurozi, R.I.M. Zen, D.P. Lestari, I. Sari, Murni (2019) Modified average of the base-level models in the hill-climbing bagged ensemble selection algorithm for credit scoring. Procedia Comput. Sci. **157**, 229–237. https://doi.org/10.1016/j.procs.2019.08.162
62. T. Hauer, Society caught in a labyrinth of algorithms: Disputes, promises, and limitations of the new order of things. Society **56**(3), 222–230 (2019). https://doi.org/10.1007/s12115-019-00358-5
63. M.L. Montagnani, C. Paulesu, Towards an ecosystem for consumer protection in the context of AI-based credit scoring. Eur. Bus. Law Rev. **33**(4), 557–580 (2022)
64. D. Tripathi, D.R. Edla, V. Kuppili, R. Dharavath, Binary BAT algorithm and RBFN based hybrid credit scoring model. Multimedia Tools Appl. **79**(43–44), 31889–31912 (2020). https://doi.org/10.1007/s11042-020-09538-6
65. S. Chakraborty, S. Aich, S.J. Seong, H.-C. Kim, A blockchain based credit analysis framework for efficient financial systems, in *International Conference on Advanced Communication Technology, ICACT, 2019*, 56–60 Feb 2019. https://doi.org/10.23919/ICACT.2019.8701926

66. S. Chava, R. Ganduri, N. Paradkar, Y. Zhang, Impact of marketplace lending on consumers' future borrowing capacities and borrowing outcomes. J. Financ. Econ. **142**(3), 1186–1208 (2021). https://doi.org/10.1016/j.jfineco.2021.06.005
67. M. Naveen Nanda, D. Venkata Subramanian, N.S. Kalyan Chakravarthy, C-score: An alternative banking methodology. Int. J. Recent Technol. Eng. **8**(2 Special), 2439–2442 (2019). https://doi.org/10.35940/ijrte.B1284.0982S1119
68. S.K. Arumugam, M. Prakash, S. Ramesh, Determinants of loan repayment behaviour of bank borrowers—A relative study with evidence from Bengaluru. Finance India **36**(1), 439–454 (2022)
69. J. Adisa, S. Ojo, P. Owolawi, A. Pretorius, S.O. Ojo, Credit score prediction using genetic algorithm—LSTM technique, in *2022 Conference on Information Communications Technology, Society, ICTAS 2022—Proceedings* (2022). https://doi.org/10.1109/ICTAS53252.2022.9744714
70. J. Hartmann, O. Hasan, A social-capital based approach to blockchain-enabled peer-to-peer lending, in *2021 3rd International Conference on Blockchain Computing and Applications, BCCA 2021*, pp. 105–110 (2021). https://doi.org/10.1109/BCCA53669.2021.9656964
71. R. Aitken, "All data is credit data": Constituting the unbanked. Compet. Chang. **21**(4), 274–300 (2017). https://doi.org/10.1177/1024529417712830
72. B. Rajendran, A.K. Pandey, V.S. Kumari Roshni, FinRep: A blockchain approach for financial reputation, in *Advances in Intelligent Systems and Computing*, vol. 1014 (2020). https://doi.org/10.1007/978-981-13-9920-6_4
73. D.P. Fonseca, P.F. Wanke, H.L. Correa, A two-stage fuzzy neural approach for credit risk assessment in a Brazilian credit card company. Appl. Soft Comput. J. **92** (2020). https://doi.org/10.1016/j.asoc.2020.106329
74. J. Ignatius, A. Hatami-Marbini, A. Rahman, L. Dhamotharan, P. Khoshnevis, A fuzzy decision support system for credit scoring. Neural Comput. Appl. **29**(10), 921–937 (2018). https://doi.org/10.1007/s00521-016-2592-1
75. N. Mehdiyev, Application of fuzzy TOPSIS for credit scoring, in *Advances in Intelligent Systems and Computing*, vol. 1029 (2020). https://doi.org/10.1007/978-3-030-23756-1_93
76. D. Xu, X. Zhang, H. Feng, Generalized fuzzy soft sets theory-based novel hybrid ensemble credit scoring model. Int. J. Financ. Econ. **24**(2), 903–921 (2019). https://doi.org/10.1002/ijfe.1698
77. Q. Li, Construction of credit system prediction model based on LM optimization algorithm, in *Lecture Notes in Electrical Engineering (LNEE)*, vol. 827 (2022). https://doi.org/10.1007/978-981-16-8052-6_171
78. T. Potkanova, L. Falat, Suggested credit score of municipalities as a tool for more efficient city management. Procedia Eng. **192**, 142–147 (2017). https://doi.org/10.1016/j.proeng.2017.06.025
79. D.F. Ahelegbey, P. Giudici, B. Hadji-Misheva, Factorial network models to improve P2P credit risk management. Front. Artif. Intell. **2** (2019). https://doi.org/10.3389/frai.2019.00008
80. G. Cheng, Financial evaluation model and algorithm based on data mining. ACM Int. Conf. Proc. Ser. 151–155 (2021). https://doi.org/10.1145/3510858.3510914
81. A. Grant, L. Deer, Consumer marketplace lending in Australia: Credit scores and loan funding success. Aust. J. Manag. **45**(4), 607–623 (2020). https://doi.org/10.1177/0312896219883678
82. M. Hossain, C. Mullally, Using evaluation data to predict loan performance among poor borrowers: The case of BRAC's asset transfer and microcredit programmes. Dev. Policy Rev. **40**(3), (2022). https://doi.org/10.1111/dpr.12579
83. H. Li, Y. Zhang, Finding the lenders of bad credit score based on the classification method. Springer Proc. Bus. Econ. 285–289 (2018). https://doi.org/10.1007/978-3-319-72745-5_31
84. H.A. Pérez, J.A. Marmolejo, J. Velasco, J.G. Fuentes, Predictive modeling approaches for payroll issuers, in *COMPSE 2016—1st EAI International Conference on Computer Science and Engineering* (2017). https://doi.org/10.4108/eai.27-2-2017.152275
85. S. Speakman, S. Sridharan, I. Markus, Three population covariate shift for mobile phone-based credit scoring, in *Proceedings of the 1st ACM SIGCAS Conference on Computing and Sustainable Societies, COMPASS 2018* (2018). https://doi.org/10.1145/3209811.3209856

86. B. Wang, J. Wang, B. Wang, R. Wang, X. Xue, Research on quantitative method of traffic safety credit score based on ridge-logistic regression, in *Lecture Notes in Electrical Engineering*, vol. 800 (2022). https://doi.org/10.1007/978-981-16-5963-8_101
87. J. Xiao, X. Zhou, Y. Zhong, L. Xie, X. Gu, D. Liu, Cost-sensitive semi-supervised selective ensemble model for customer credit scoring. Knowl. Based Syst. **189** (2020). https://doi.org/10.1016/j.knosys.2019.105118
88. A.V. Ertemel, G. Çaylak, The effect of personality traits on credit score using Myers-Briggs Type Indicator (MBTI) personality types, in *Eurasian Studies in Business and Economics*, vol. 19 (Springer Science and Business Media, 2021), pp. 185–199. https://doi.org/10.1007/978-3-030-77438-7_11
89. F.G. Mohammadi, M.H. Amini, H.R. Arabnia, Applications of nature-inspired algorithms for dimension reduction: Enabling efficient data analytics, in *Advances in Intelligent Systems and Computing*, vol. 1123 (Springer, 2020), pp. 67–84. https://doi.org/10.1007/978-3-030-34094-0_4
90. A. Soroka, G. Bristow, M. Naim, L. Purvis, Measuring regional business resilience. Reg. Stud. **54**(6), 838–850 (2020). https://doi.org/10.1080/00343404.2019.1652893
91. L. Alfat, M. Rizkinia, R.F. Sari, D.M. Romano, Feature selection of credit score factor based on smartphone usage using MCFS, in *TIMES-iCON 2019—2019 4th Technology Innovation Management and Engineering Science International Conference* (2019). https://doi.org/10.1109/TIMES-iCON47539.2019.9024462
92. P.K. Jadwal, S. Jain, B. Agarwal, Spectral clustering and cost-sensitive deep neural network-based undersampling approach for P2P lending data. Int. J. Inf. Technol. Web. Eng. **15**(4), 37–52 (2020). https://doi.org/10.4018/IJITWE.2020100103
93. A.F. Borthick, G.P. Schneider, T.R. Viscelli, Analyzing data for decision making: Integrating spreadsheet modeling and database querying. Issues Account. Educ. **32**(1), 59–66 (2017). https://doi.org/10.2308/iace-51385
94. K. Panyagometh, Impact of baseline population on creditscore's predictive power. Econ. Sociol. **12**(1), 262–269 (2019). https://doi.org/10.14254/2071-789X.2019/12-1/15
95. A. Senthil Kumar, S. Ramesh, S. Rahul, A technology on credit score system-assessing public perception in Bengaluru city. Int. J. Innovative Technol. Exploring Eng. **8**(12), 1929–1934 (2019). https://doi.org/10.35940/ijitee.L2890.1081219
96. S. Lee, Determining personal credit rating through voice analysis: Case of P2P loan borrowers. KSII Trans. Internet Inf. Syst. **15**(10), 3627–3641 (2021). https://doi.org/10.3837/tiis.2021.10.009
97. H. Ots, I. Liiv, D. Tur, Mobile phone usage data for credit scoring **1243 CCIS**(February 2020), 82–95 (2020). https://doi.org/10.1007/978-3-030-57672-1_7
98. N. Torvekar, P.S. Game, Predict. Anal. Credit. Score Credit. Card Defaulters **7**(5), 283–286 (2019)
99. H.R.A. Farid Ghareh Mohammadi, M. Hadi Amini, An introduction to advanced machine learning: meta-learning algorithms, applications, and promises, in *Advances in Intelligent Systems and Computing*, p. 135 (2020). http://www.springer.com/series/11156
100. V. Hassija, G. Bansal, V. Chamola, N. Kumar, M. Guizani, Secur. Lend.: Blockchain Prospect. Theory-Based Decentralized Credit. Scoring Model. **7**(4), 2566–2575 (2020). https://doi.org/10.1109/TNSE.2020.2982488
101. A.H. Rahat et al. Blockchain based secured multipurpose identity (SMID) management system for smart cities, in *2022 IEEE 12th Annual Computing and Communication Workshop and Conference, CCWC 2022*, pp. 737–744 (2022). https://doi.org/10.1109/CCWC54503.2022.9720808
102. N.U. Khan, Q.U.A. Shah Jehan, A. Shah, Impact of taxation on dividend policy: evidence from Pakistan **42**, 365–375 (2017). https://doi.org/10.1016/j.ribaf.2017.07.157
103. A. Ashofteh, J.M. Bravo, A conservative approach for online credit scoring. Expert Syst. Appl. **176** (2021). https://doi.org/10.1016/j.eswa.2021.114835
104. F. Ciampi, V. Cillo, F. Fiano, Combining Kohonen maps and prior payment behavior for small enterprise default prediction. Small Bus. Econ. **54**(4), 1007–1039 (2020). https://doi.org/10.1007/s11187-018-0117-2

105. C. Liu, H. Huang, S. Lu, Research on personal credit scoring model based on artificial intelligence, in *Advances in Intelligent Systems and Computing*, vol. 929 (2019). https://doi.org/10.1007/978-3-030-15740-1_64
106. J. Prassanna, A.Q. Md, J. Christy Jackson, R. Prabakaran, S. Ramanathan, SecrecyProtector: A novel data analytics based credit score management system. Int. J. Sci. Technol. Res. **9**(3), 1215–1218 (2020)

Awareness of COLREGS Among Fishermen in Malaysia

Silloshina Manoharan and Mohammed Ismail Rustam Suhrab

1 Introduction

This paper describes the study of the level of awareness of fishermen toward COLREGs that will include a background, statement of problems, objectives, methodology, discussion, data collection, data analysis, and conclusion. This title was inspired by a chance observation of fishermen in the Kuala Terengganu area who seemed to disregard the COLREGS while at sea. International Regulations and Prevention Collisions (COLREGs) is a law established by the International Maritime Organization (IMO) in 1972, which has established navigation rules that must be followed by vehicles at sea to avoid collisions between ships [1]. When a vessel is underway, the COLREGS apply to it. Importantly, fishing vessels must adhere to the COLREGS when they are engaged in fishing, and their attention is diverted elsewhere [2]. Everyone must respect the "rules of the road" at sea, whether on the largest ship or a fishing boat three miles off the shore. With the implementation of this study, the knowledge gathered during the study will be able to be improved.

Background

Ships transport a wide range of products to and from Malaysian ports, which are vital to the country's economy [3]. Unfortunately, ships and fishing vessels collide often. According to Shaiful, most fishermen today are unaware of COLREGs. This is due to

S. Manoharan · M. I. R. Suhrab (✉)
Faculty of Maritime Studies, University Malaysia Terengganu, Kuala Terengganu, Malaysia
e-mail: m.ismail@umt.edu.my

S. Manoharan
e-mail: s54558@ocean.umt.edu.my

A. Fitriadhy et al. (eds.), *Advancing Interdisciplinary Knowledge and Innovation*,
SpringerBriefs in Applied Sciences and Technology,
https://doi.org/10.1007/978-3-032-01309-5_4

several mishaps resulting from a ship or boat failing to comply with the COLREGs. The fishing vessel does not always do the right thing in the lead-up to the disaster. The fishermen contributed to the collision by failing to meet their responsibilities [4]. The concept for this study arose as the number of maritime accidents among fishermen increased. This study was done to determine the level of knowledge among fishermen to indirectly enhance their understanding of COLREGs. This study will also look at some of the common variables that contribute to collisions in the hope that fishermen will become aware of the issues and make efforts to mitigate the consequences of a collision.

Objective

- To examine the level of awareness of fishermen against COLREGs.
- This study was conducted to show fishermen in the Kuala Terengganu area the importance of complying with COLREGS.
- To recommend fishermen in the coastal area of Kuala Terengganu to practice the right law despite being accustomed to marine conditions and the environment.

2 Methodology

To finish this study, both primary and secondary data were gathered. However, since it is a quantitative study, data from both primary and secondary sources must be gathered. The use of a questionnaire is one way to obtain primary data. A questionnaire is a type of research tool that asks respondents a series of questions to collect data. In contrast to other survey methods, such as verbal or telephone surveys, a questionnaire is less expensive and time-consuming for the researcher. Additionally, questions typically include standardized responses that make compiling the data received quick and easy.

The questionnaires are based on the study's goals to determine whether fishermen know collision regulations. The researcher had to investigate the causes of fishing vessel accidents at sea before developing the questionnaire. Fishermen in the Kuala Terengganu region were given the prepared questionnaire. The underlying causes of fishing vessel accidents were determined by analyzing information and utilizing SPSS. The opinions and ideas of the respondents on the awareness of COLREGS were gathered through interviews and surveys that were fully completed. The interviewees were each given a 30-to-40-min interview.

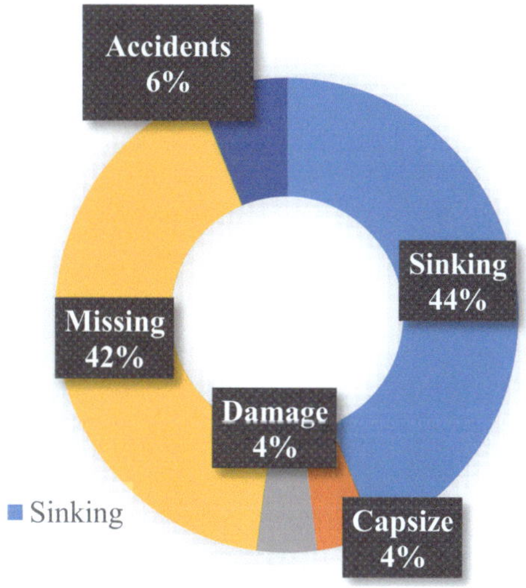

Fig. 1 Incidents involving ships and boats 2020

The combination of software applications bundled into a single package is known as Statistical Package for the Social Sciences (SPSS). This program's primary use is to analyze scientific information about social science. Pulling, modifying, and analyzing data are clean and simple tasks when they are done in SAV format. Researchers' work is significantly simplified by SPSS, which uses the SAV format to automatically build up and import the desired variable names, variable types, titles, and value labels. Survey data can be exported to SPSS, where there are countless chances to conduct statistical analysis.

The researcher had to read various literature reviews and study-related papers to determine the awareness of COLREGS among fishermen. According to data from the Malaysian Maritime Enforcement Agency [5, 6], 1267 fishing vessels were accused of breaking the law in Malaysian waters in 2020. Further research is conducted to determine the level of awareness about COLREGs among Kuala Terengganu fishermen. In terms of data collection, a questionnaire is being prepared and produced, and a survey will be conducted among the fishing community. This will be useful in proposing a solution to reduce collisions or occurrences involving fishing vessels. Kindly refer to Figs. 1 and 2.

3 Discussion

This study reveals the issues that could arise if fishermen do not adhere to COLREGs at sea. However, there are significant issues with the results of this study, particularly with the information gathering about fishing-related incidents.

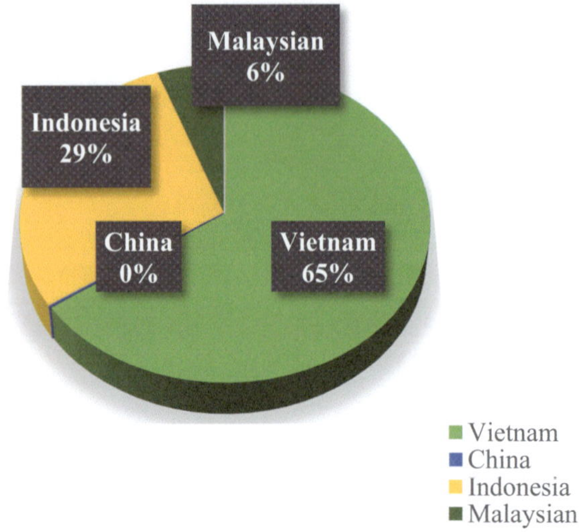

Fig. 2 Number of fishing vessels suspected of violating the law in the Malaysian sea 2020

The goal is to raise knowledge of COLREGs among fishermen in the Kuala Terengganu area, which will assist in preventing fishing-related accidents at sea. The study indirectly educated fishermen on the significance of COLREG knowledge. As a result, the study's objectives have been met. The data reported in this chapter show that fishermen should be familiar with and use COLREGs.

4 Research Analysis

These findings are related to the research topics at this study's heart. Data were evaluated to identify and depict fishermen's awareness of COLREGs in the Kuala Terengganu area. In this study, we may learn about the issues that may arise if fishermen do not use COLREGs at sea. Accidents involving fishermen and other modes of water transport are examples, as is pollution caused by oil spills. However, there are significant issues with the study's conclusions, particularly in collecting information on fishermen-related accidents. The data for this case study came from a questionnaire distributed to thirty fishermen in the Kuala Terengganu area. A total of thirty questionnaires were received for this study, all of which matched the requirements outlined in the previous chapter. Refer to Table 1.

Table 1 Analysis of questions and answers

No	Question	Yes	No
1	Is your boat registered with the Department of Fisheries? Malaysia/other departments?	30	0
2	Is your boat license still active?	27	3
3	Do you know about the International Regulations *Preventing Collisions at Sea (COLREGs)?*	27	3
4	Do you apply COLREGs while at sea?	30	0
5	Do you have a rank such as Master/Ship Owner/Crew? (Rule 2 COLREGs—Responsibility)	24	6
6	Does every rank/position on the ship/boat fulfill their responsibilities as enshrined in the law of the ship?	24	6
7	Does your boat have communication tools?	30	0
8	Does your boat have emergency equipment such as Safety Jackets, Lifebuoys, and Emergency Boxes?	30	0
9	While at sea, did you receive information/data on the position of other ships?	18	12
10	Did you receive information on weather and sea conditions before/while at sea?	27	3
11	Do you know the types of lights (night) and symbols (day) for your boat type? (Rules 22 COLREGs—Visibility of Lights)	27	3
12	Are the position of the lights and symbols for your boat type according to COLREGs? (Rules 22 COLREGs—Visibility of Lights)	24	6
13	Do you use other types of lights and symbols?	15	15
14	To your knowledge, does fishing have a set area limit?	30	0
15	Do you comply with the area limits?	27	3
16	Do you know and adhere to the anchorage area?	24	6
17	Do you know the areas to avoid catching fish?	24	6
18	Do you dispose of garbage in the sea area according to the laws of the sea?	15	15
19	During an emergency, do you give instructions according to COLREGs?	12	18
20	In your opinion, can fishermen who apply COLREGs help reduce accidents at sea?	27	3
21	In your opinion, do fishermen need to know about COLREGs?	27	3

5 Conclusion

This study has the potential to enhance awareness of the importance of COLREG among fishermen around the Kuala Terengganu coastline and community (International Regulations and Preventing Collision at Sea). The case study's data was collected from a survey given to thirty fishermen in the Kuala Terengganu region, as shown in Table 1. This study can be utilized as a resource for fishermen who are believed to be unaware of COLREG. With the completion of this research, the younger generation, who are new to the science of navigation, will be able to follow the regulations that should be followed for both fishing boats and huge vessels. According to the study, most fishermen have taken courses to earn a fishing license. They were exposed to little about COLREGs during the training, but a few fishermen still did not know what they were. This is because the crews are not required to follow the course. As a result, we urge people who wish to continue this research to investigate the extent of COLREG adoption among fishermen. The data reported in this article show that fishermen should be familiar with and use COLREGs.

Acknowledgements The authors would like to thank the Faculty of Maritime Studies, University Malaysia Terengganu, and everyone who helped with this study.

References

1. IMO, Convention on the International Regulations for Preventing Collisions at Sea, 1972 (COLREGs) (2019). https://www.imo.org/en/About/Conventions/Pages/COLREG.aspx Accessed 24 May 2023
2. S.B. Ismail, S.E.A. Hamid, Development of safety modelling based on human factor for fishermen zone a: a case study At Minjung's District **11**(2), 70–79 (2020). https://www.mimet.edu.my/v07/index.php/research-and-innovation/technical-bulletin/category/9-marine-frontier-vol-11-ed-2-2020 Accessed 22 May 2023
3. Kurangkan risiko nahas di laut, Sinar online (2015). http://www.sinarharian.com.my/edisi/pahang/kurangkan-risiko-nahas-di-laut-1.445863. Accessed May 2023
4. D. Jin, E. Thunberg, An analysis of fishing vessel accidents in fishing areas off the northeastern United States (2005). https://www.researchgate.net/publication/222648103_An_analysis_of_fishing_vessel_accidents_in_fishing_areas_off_the_northeastern_United_States. Accessed 22 May 2023
5. Malaysian Maritime Enforcement Agency annual report, p. 42 (2020). https://www.mmea.gov.my/artikel/Laporan_Tahunan/APMM_LaporanTahunan2020_WEB.pdf. Accessed May 2023
6. Malaysian Maritime Enforcement Agency annual report (2019). https://www.mmea.gov.my/artikel/Laporan_Tahunan/APMM_AnnualReport2019.pdf. Accessed May 2023

Risk Assessment Study on Ship Accidents at Port of Tanjung Pelepas

Nurul Najwa Nursyafrinna Mahzil, Muhamad Nasir Rahmatdin, and Mohammed Ismail Russtam Suhrab

1 Introduction

This paragraph entails the introduction of the study, which captures the history of the research area and the background of the study. The problem statement that would elaborate on a few problems in this research area, research question, and objectives will be described that can be used further to minimize or reduce the accident risk at port of Tanjung Pelepas [1].

Ports are a vital economic activity intended to provide facilities for interchange cargo and passenger traffic among vessels and vessels to and from inland locations by rail, road, inland waterways, and pipeline. A port can be a harbor or area supporting many boats and vessels transferring people or cargo while allowing continuous or periodic shipment transactions [2]. The potential source of accidents in port operations, for instance, spills, explosions, and fires are closely linked to the function of the port itself and to the installations and activities associated with it, which is its transfer from water to land and vice versa of large amounts of waterborne cargo with a wide diversity of materials such as oils, coal, chemicals, and cars [3]. On the other hand, activities at a harbor involve additional hazards, such as loading and unloading materials to or from ships, and oil and chemical tankers, and navigating and maneuvering in a restricted area and oil jetties [3, 4].

As a result of this situation, accidents occur in ports with a certain frequency, including some with several consequences. An investigation to identify the factors

N. N. N. Mahzil · M. N. Rahmatdin · M. I. R. Suhrab (✉)
Faculty of Maritime Studies, Universiti Malaysia Terengganu, Kuala Terengganu, Malaysia
e-mail: m.ismail@umt.edu.my

N. N. N. Mahzil
e-mail: s55304@ocean.umt.edu.my

M. N. Rahmatdin
e-mail: muhamadnasir@umt.edu.my

A. Fitriadhy et al. (eds.), *Advancing Interdisciplinary Knowledge and Innovation*,
SpringerBriefs in Applied Sciences and Technology,
https://doi.org/10.1007/978-3-032-01309-5_5

that cause ship accidents at port of Tanjung Pelepas would be remarkably interesting. Is the frequency of accidents at port of Tanjung Pelepas increasing? What are the most common factors that cause accidents? What risk factors or causes of ship accidents have been identified? Can these accidents minimize the risk at port of Tanjung Pelepas with any recommendations [5]?

Identifying the risk and its origins and the occurrence and severity of the risk's effects are part of the risk assessment process. This is required to develop risk reduction and safety improvement plans at sea by implementing risk prevention, control, and reduction methods [2]. Risk includes damage and uncertainty, to say, the possibility of loss or injury and the degree of probability of such loss. Navigational risk can be assessed based on the frequency, consequences, and impacts of accidents in a specific area. It also needs to implement the process to reduce risk using the standard method. In studying marine accidents regarding navigation, most of the accidents occurred in a restricted area, in the port limit area, or under pilotage operation. Therefore, it can be minimized as much as possible to avoid a major accident within the limit of port of Tanjung Pelepas.

2 Methodology

There is no specific record of the ship accidents at PTP. So, this study will assess the safety at the port of Tanjung Pelepas through selected experts at PTP. In this research, several methods are used to fulfill the objectives. The data obtained are from primary and secondary sources, including interviews, surveys, analytical data, and observations. This paragraph of the research shows the methodologies, techniques, and how to collect the data, as shown in Fig. 1.

The first methodology used in this study to achieve the first objective is hazard identification (HAZID). Site observation and interviews have been conducted to identify the possible hazards.

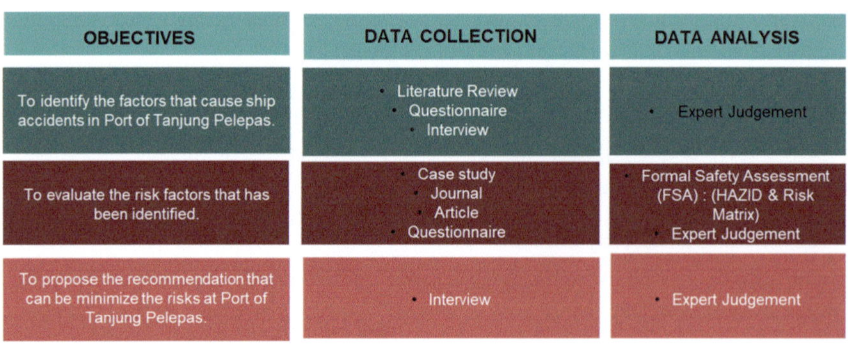

Fig. 1 Research design

The second methodology used to achieve the second objective is the Risk Matrix method. The departments authorized to regulate safety at port of Tanjung Pelepas were given a questionnaire to obtain the data for completing the objectives of this research.

Representatives of departments at port of Tanjung Pelepas were randomly selected for interviews to obtain data and information on the dangers or risks found at PTP. The data obtained from the questionnaire is analyzed and evaluated by using the risk matrix methodology.

3 Results and Discussion

The first objective was identifying the factors that cause ship accidents at port of Tanjung Pelepas and enhancing hazard identification, literature review, and questionnaire. From the observation and site survey, this port has eight main hazards: contact with a jetty at inner harbor quays, mooring hazards, transferring cargo between ships and wharf, falling objects, uncovered drains, hazardous substances, moving vehicles and equipment, and slips and trips [6]. The risk details were also collected from all these hazards, such as hazard description, type of hazardous situation, type of potential risk, the severity of consequences, and probability of an incident. The second objective was to rank the hazards using a risk matrix to find the highest frequency and consequence of hazards and the risk percentage at the port of Tanjung Pelepas. Then, the recommendation is made to control the risk, and the area of improvement can be achieved.

Hazard Identification

Table 1 shows the hazards and risks at the port of Tanjung Pelepas that have been listed and verified.

HAZID 1: Inner Harbor Quays

The Port of Tanjung Pelepas has more than three terminals, Malaysia's most prominent port with a technologically advanced container terminal [7]. As there are more than three terminals, hazards at inner harbor quays exist, such as the ship contacting the jetty during berthing operation. This hazard is one of the highest risks in port operations. The risk matrix analyzed indicates a moderate risk matrix affecting property, environment, and stakeholders. Hence, efforts should be made to reduce the risk to "As low as reasonably practicable" (ALARP), but activity may be undertaken [8].

However, the people category is minor; hence, no additional controls are required, but monitoring is required to ensure no circumstances change. Therefore, the recommendations that have been discussed with the port authority during the interview session to reduce risk during berthing operations are as follows:

Table 1 List of hazards at PTP

Activity	Hazard	Risk
Berthing	1. Inner harbor quays: The ship contacts the jetty or quay during the berthing operation	The ship collides with a berth
	2. Mooring hazards: Failure of ship moorings/land-based mooring equipment	Trapping hands in mooring ropes
Cargo operations	3. Transferring cargo between ships and the wharves: The incident occurred during loading/unloading operations due to equipment failure or human error	Dropped load leads to damage to the ship
	4. Falling objects: The incident occurred during the stacking and stowing of goods	Dropped/Falling load leads to structural damage to the ship
	5. Hazardous substances: Workers loading and unloading solid bulk cargoes exposed to dust or respiratory sensitizers and cargoes may produce gases	It can cause asthma and difficulty in breathing
	6. Moving vehicles and equipment: The traffic management system is not appropriate and standard	Fatality Accident or significant injury

- A maintenance program for navigation aids in the shipping channel.
- Maintenance of an adequate shipping channel (width and depth).
- Use of tugs.

HAZID 2: Mooring Hazards

During berthing, persons engaged in mooring and unmooring operations will start their duty. Mooring can be hazardous as a person is at risk of getting caught in a line. The lines can be heavy and awkward, particularly if wet, and may break and snap back. Based on the interview session with the operations officer, the cause of failure of the mooring was the rotted ropes.

According to the risk matrix that has been analyzed, the risk matrix that affects people is moderate; hence, efforts should be made to reduce the risk to "As low as reasonably practicable" (ALARP). Therefore, recommendations have been made after the discussion with the port authorities. All personnel engaged in mooring operations should get advanced training [9]. They should.

- Wear the proper personal protection equipment.
- Keep a safe distance from being hit by lines and be prepared to pick them up as soon as they land.
- Move into the snap-back zone when it is necessary to do so.
- Never step over a mooring line under tension or place their feet in the bights or eyes of mooring lines.

HAZID 3: Transferring Cargo Between Ships and Wharves

As an all-weather port, the port of Tanjung Pelepas can handle vessels of up to five million TEUs and several types of cargo, such as dry and liquid bulk [10]. It is also a regional center for transshipment, cargo consolidation, and distribution activities. Therefore, when transferring cargo between ships and wharves, the hazard may exist when equipment failure or human error occurs. Cargo will drop and lead to damage to ships and personnel.

The risk matrix table indicates that the risk matrix that affects people, property, and stakeholders is moderate; hence, efforts should be made to reduce risk to "As low as reasonably practicable" (ALARP), but activity may be undertaken. However, the risk matrix table indicates that this hazard is at a minor stage for the environmental aspect [11]. Hence, no additional controls will be required, but monitoring is required to ensure no circumstances change. Here are the possible recommendations that can reduce the risk during port operations:

- A Loading and Unloading Management Plan must be implemented.
- Emergency Response Plans must be provided.

HAZID 4: Falling Objects

There is a risk of falling objects when loading and unloading activities and stacking and stowing goods [12]. When loads of things become unstable during transport or are improperly loaded, they may collapse or fall. Items could be loose and slung, stacked, or slung wrongly. It is possible to drop fixtures and fittings utilized during lashing operations. Based on the risk matrix table, the level of risk that affects people is moderate; hence, efforts should be made to reduce the risk to "As low as reasonably practicable" (ALARP), but activity may be undertaken.

For the property category, the level of risk is at the minor stage. Hence, no additional controls are required, but monitoring is required to ensure no circumstances change. As for the environment and stakeholder category, the level of risk is at a slight stage. Hence, there will be no action required as there is an insignificant impact on the environment and stakeholders' business [13]. Therefore, this is the way to reduce risk during stacking and stowing goods:

- Safe work systems must be in place to ensure that pre-slung and lost loads can be lifted safely.
- Every piece of security hardware, including twist locks and lashing bars, must be thoroughly inspected.
- Loads must be properly secured, particularly when moving around the dock.
- In container ports, marked safe spaces for lorry drivers to maneuver during loading and unloading operations should be available.
- Designated safe places should be made available for customs agents to inspect and seal containers.
- Good ship-to-shore contact and coordination are necessary to load and unload substantial and stiff bulk commodities.

HAZID 5: Hazardous Substances

Cargoes could be caustic, toxic, explosive, or poisonous. Even though some solid bulk cargoes in the hold, like fishmeal or bark, may not be harmful, they could nonetheless release gases due to bacterial action or decomposition. Some cargoes, including grain, may have undergone fumigation. Hazardous vapors could be produced by vehicle exhaust emissions in the ship's hold [14]. Dust or respiratory sensitizers that can trigger asthma may be exposed to workers loading and unloading solid bulk freight [15].

The table indicates that the risk matrix that affects people is moderate; hence, efforts should be made to reduce the risk to "As low as reasonably practicable" (ALARP), but activity may be undertaken. The table for the property and stakeholders' categories indicates that this hazard has an unlikely frequency, and the consequences are insignificant. It is shown that the level of the risk matrix is light. Hence, there will be no action required as there is no impact on property and stakeholders' business. These are the recommendations for reducing the risk of hazards:

- When necessary, closed systems should be used to prevent dust escape.
- Grabs must be maintained following safe work standards to avoid dust creation.
- The vehicle exhaust emissions in a ship's hold need to be measured.
- Safe work procedures must be followed before entering the cargo hold.
- A confined space should not be entered unless it has been ventilated and a qualified individual has given it the all-clear.

HAZID 6: Moving Vehicles and Equipment

The observation during the site visit shows that vehicle and pedestrian routes are not segregated. An inappropriate traffic management system may lead to danger and risk if no control measure is set up [16]. The potential risk that may exist caused by this hazard is accident and loss of life [17]. Based on the table of risk matrices that have been analyzed, these hazards are also one of the highest risks in this port. The table indicates that this hazard has a likely frequency, and the consequences to people are at a major level. Therefore, it is shown that the level of the risk matrix is at an extreme stage, which is the highest risk. Hence, it results in intolerable risks, and this activity is not authorized, and recommendations must be suggested to reduce the risk.

However, no additional controls for the property, environment, and stakeholders' categories will be required. Still, monitoring is required to ensure no changes in circumstances, as the level of risk affecting these three categories is minor. Therefore, these are the recommendations to reduce the risk:

- The port's operational management and safety will be aided by implementing an effective traffic management system.
- Drivers should be informed by signage on which side of the road they should use.
- Designated spaces for pedestrians and vehicles should be placed.
- Access to operational areas should be prohibited for public and private automobiles, taxis, and delivery vehicles.

4 Conclusion

Risk assessment is important in ensuring a safe work environment at the port. Every stakeholder in port areas needs to be aware of hazard sources and risk factors that could cause harm. The objectives of this research project were focused on identifying hazards, risks of hazards, and giving workable solutions and recommendations to improve the safety of port operations at port of Tanjung Pelepas.

Respondents involved in this research are from the Port operations department, marine department, and safety department of PTP. Based on the first objective, hazards at the port of Tanjung Pelepas were identified and verified during the site observations and interviewed with the officers from the three departments. Using the HAZID method, it can be concluded that six hazards have the potential to cause harm to people, the environment, property, and harbor stakeholders.

Based on the second objective, an assessment that had been conducted successfully ranked the most potential hazards that may contribute to the elevated risk of accidents at the port of Tanjung Pelepas. The hazards were calculated using the risk matrix method by frequency and consequences. Each hazard has its level of frequency and consequences. The most potential hazards at the port of Tanjung Pelepas are inner harbor quays where ships come into contact with jetty during berthing and moving vehicles and equipment. Both hazards indicate the highest number of the matrix index, while the lowest risk of hazards is falling objects during stacking and stowing goods.

As for the third objective of this research, some workable solutions have been proposed and discussed to reduce the risk of accidents and improve the safety level during port operations. To summarize, recommendations have been made to enhance berthing and cargo operations at the port. Such recommendations include the berthing category's inner harbor quays and mooring hazards. Meanwhile, other recommendations for cargo operations have been made for falling objects, hazardous substances, moving vehicles, and equipment. It can also be concluded that most of the hazards in PTP are controllable and can be reduced. This research has contributed to improving safety at the port of Tanjung Pelepas.

Acknowledgements We acknowledge all the officers at the port of Tanjung Pelepas who were involved and cooperated best during the data collection in the interview session. Besides, we are thankful to the Faculty of Maritime Studies, Research and Field Service Centers of University Malaysia Terengganu for allowing us to conduct this study.

References

1. PTP. Pelabuhan Tanjung Pelepas Sdn Bhd (2020). https://www.ptp.com.my/. Accessed May 2023
2. D. Soumyajit, What Are the Different Types of Ports for Ships? Marine Insight (2020). https://www.marineinsight.com/ports/what-are-the-various-types-of-ports/ Accessed May 2023

3. T. Pelepas, Port of Tanjung Pelepas Terminals (2019) https://www.apmterminals.com/en/tanjung-pelepas/about/our-terminal. Accessed 25 May 2023
4. M.F. Shah, R. Nordin, *Port of Tanjung Pelepas jumps to 15th in world ranking | The Star*, pp. 1–7 (2021) https://www.thestar.com.my/business/business-news/2021/08/04/port-of-tanjung-pelepas-jumps-to-15th-in-world-ranking. Accessed August 2023
5. C. Management, M. Line, C. List, W. Blower, S. Chain, *World Top Container Ports 2012*, pp. 1–2 (2012)
6. M. Dg, *In a statement this evening, the Singapore Maritime and Port Authority claimed that the incident occurred in Singapore territorial waters within Singapore port limits off Tuas*, pp. 11–13. (2019). https://www.malaysiakini.com/news/463344. Accessed June 2023
7. M. Mansor, A.S. Abd Kader, M.R. Jalal, Risk assessment of marine accidents in Tanjung Pelepas port limit due to reclamation of Tanjung Piai maritime industrial. Jurnal Mekanikal June 1–13 (2019)
8. M'sian gov't vessel collides with Greek ship in disputed waters. Malaysiakini News, 1–3 (2019) https://www.malaysiakini.com/news/463334. Accessed May 2023
9. O.F. Knudsen, B. Hassler, IMO legislation and its implementation: accident risk, vessel deficiencies, and national administrative practices. Mar. Policy **35**(2), 201–207 (2011). https://doi.org/10.1016/j.marpol.2010.09.006
10. CCOHS. What is a hazard, and what is the risk? Canadian Centre for Occupational Health and Safety, March 1–8 (2020) http://www.dmp.wa.gov.au/Safety/What-is-a-hazard-and-what-is-4721.aspx. Accessed June 2023
11. N.M. Parra, A. Nagi, W. Kersten, Risk assessment methods in seaports—a literature review, in *HAZARD Project* (vol. 24, issue May) (2018) https://doi.org/10.13140/RG.2.2.32359.50081
12. T.O.F. Contents, What Does Accident Rate Mean? 1–14 (2021)
13. M. Hurley, E. Rosenbaum, Hazard and risk. Perform.-Based Fire Saf. Des. 23–36 (2015). https://doi.org/10.1201/b18375-3
14. Formal safety assessment. IMO International Maritime Organization (2019). https://doi.org/10.4324/978080473369-19
15. Y. Bai, W.-L. Jin, Formal safety assessment applied to shipping industry. Mar. Struct. Des. 765–780 (2016). https://doi.org/10.1016/b978-0-08-099997-5.00041-1
16. Sunaryo, A. Lubis, M.B. Ramadhan, Increasing the safety of standard shipping using a formal safety assessment approach, in *AIP Conference Proceedings*, vol 2227 (2020). https://doi.org/10.1063/5.0001027
17. C. Gasparotti, E. Rusu, Methods for the risk assessment in maritime transportation in the black sea Basin. J. Environ. Prot. Ecol. **13**(3 A), 1751–1759 (2012)

Designing Quantitative and Qualitative Assessment for Walkability and Bikeability Within the Railway Environment in Greater Jakarta

Widyanesti Liritantri, Widyarini Weningtyas, Kardina Nawassa Setyo Ayuningtyas, and Anisah Pratiwi

1 Introduction

Providing accessible public transportation is one of the targets of public policies and urban planners around the world; nevertheless, many places have low level of accessibility, and this is the reason why people do not use public transportation [1]. The improvement of accessibility becomes a priority to alleviate several aspects, including social, economics, and environmental [2]. Moreover, active transport has many positive impacts on health, quality of life, and environment [3]. A sustainable approach to improving accessibility is to encourage active mobility, namely by walking and cycling. Walking and cycling have both individual and collective benefits, such as preventing climate change and air pollution, improving public health and mental health, and individual and financial benefits [4, 5]. Walkability indicators are a pivotal method to evaluate the role of the built environment regarding active mobility, and to promote more sustainable and equitable regions [6].

Investigations to assess the accessibility of walking and cycling have been carried out in several studies. Assessment of pedestrian accessibility at train stations has been carried out previously based on public transport accessibility levels (PTALs),

W. Liritantri
Telkom University, Jl. Telekomunikasi Bandung, Bandung, Indonesia
e-mail: widyanesti@telkomuniversity.ac.id

W. Weningtyas (✉) · K. N. S. Ayuningtyas · A. Pratiwi
Institut Teknologi Bandung, Jl Ganesha 10, Bandung, Indonesia
e-mail: widyariniw@itb.ac.id

K. N. S. Ayuningtyas
e-mail: kardina.nawassa@itb.ac.id

A. Pratiwi
e-mail: pratiwianisah12@gmail.com

A. Fitriadhy et al. (eds.), *Advancing Interdisciplinary Knowledge and Innovation*,
SpringerBriefs in Applied Sciences and Technology,
https://doi.org/10.1007/978-3-032-01309-5_6

by calculating walking time and availability of transportation services [7]. Another study measured walking accessibility in East Java based on the correlation of land use factors, service quality at stations, walking distance, and walking time [8]. This study shows the need to improve the quality of access to and from stations.

The average Indonesian walks only 3513 steps per day, compared to Hong Kong, Singaporean, and the average British person who walks more than 5000 steps per day. "In more walkable cities, activity was greater throughout the day and throughout the week, across age, sex, and body mass index (BMI) groups, with the greatest increase in activity for women" [9].

The issue of walking accessibility in public transport is about how easy it is to access the public transport system and how to reach the destination from the terminal on foot. The possibility of walking is often associated with proximity to the destination. Furthermore, several aspects are barriers for people to walk, including travel distance/time, security and safety, weather, comfort, and facilities [10, 11]. Other studies have shown that other factors such as the availability of information facilities also play a role in station accessibility. People who have been lost or confused about accessing stations tend to score lower accessibility scores [12]. Active transport has lost its importance and should be prioritized by stakeholders [13]. Furthermore, the value of accessibility is relative. The infrastructure must be able to allow everyone to access the station easily and independently. Therefore, this aims to formulate walkability and bikeability assessment methodology determining selected rail station's location, designing a quantitative and qualitative survey using a visual illustration.

2 Methodology

In this study, we consider using a radius of 400 m from the station to check the walkability and bikeability measurements. After final assessment has been carried out, we can determine the catchment area for each station.

The research framework was based on a wide scope of actual conditions (such as demographic, surrounding region mobility, economic activity, air quality, and private vehicle ownership). From the actual conditions, we narrow down the problems which are the accessibility to rail transit station, and as a solution to our study objectives are walkability and bikeability assessment and improvements. There are three methods considered in measuring walkability, namely PLOS, equivalent walking distance (EWD), and LOA.

1. PLOS—quantitative and qualitative measurement: The highway capacity manual (HCM) 2010 method calculates the pedestrian flow rate (Vp) and average pedestrian space (Ap) in assessing the level of service (LOS) and walking accessibility. The higher the volume and the density of pedestrians, the speed will decrease. The higher the density and the lower the space, the level of mobility will decrease, just like the speed of pedestrians:

$$Vp\,(pedestrian/ft/min) = \frac{pedestrian\,flow_{15\,min}}{15\,x\,effective\ walkway\ width(ft)} \quad (1)$$

$$The\ average\ pedestrian\ space(Ap) = \frac{pedestrian\ speed\,(Sp)}{pedestrian\,flow_{15\,min}(V_{15})} \quad (2)$$

$$PLOS_{(TQM)} = Avaraged\ Score = \frac{C_1 + C_2 + C_3 + C_4 + \cdots + C_n}{n} \quad (3)$$

2. LOA—qualitative measurement: LOA (Permen PU 03/PRT/ M/2014) assignment is based on the perspective of the passenger trains which considers several parameters such as easiness, convenience, and safety.
3. Bikeability assessment (HCM 2010 and AASTHO 2012)—BLOS: BLOS is the most accurate method of evaluating bicycle conditions of shared roadway environment:

$$BLOS = 0.760 + Fv + Fs + Fp + Fw \quad (4)$$

$$Fv = 0{,}507\,ln\left(\frac{Vma}{4\,Nth}\right);$$

$$Fs = 0{,}199[1{,}1199ln(Sra - 20) + 0{,}8103(1 + 10{,}38PHva)^2] \quad (5)$$

$$Fp = 7{,}066\left(\frac{1}{Pc}\right)^2;\ Fw = -0{,}005 - We^2 \quad (6)$$

Research Locations

Greater Jakarta (including Jakarta, Bogor, Depok, Tangerang, and Bekasi) has 126 railway stations, which consist of 89 commuter line stations, 13 MRT stations, 6 LRT stations (Jakarta Line), and 18 LRT stations (JABODETABEK Line—under construction). The research locations were selected by 4 phases of eliminations (Table 1).

Designing the Qualitative Assessment by Using Visual Illustrations

Qualitative assessments elaborate on respondents' perception about the pedestrian facilities. Therefore, we will need some visual photos or illustrations in order to

Table 1 List of stations for research location

Type of stations	Number of stations	Selected station	Elimination phase			
			Phase I identify type of stations	Phase II integration system and land use	Phase II pedestrian (P) and cycling (B)	Phase IV development plan
Commuter Line Station	24	Sudirman	V	V	√P √B	V
		Tanah Abang	V	V	√P√B	V
		Manggarai	V	V	√P	V
		Cawang	V	V	√P	V
		UI	V	V	√P√B	
		Juanda Station	V	V	√P√B	
		Bekasi	V	V	√P	
		Bogor	V	V	√P√B	V
MRT	13	13 stations	V	V	√P√B	V
LRT	6	6 stations	V	V	√P	V

elaborate on the pedestrian condition and receive respondents' perception in a same manner. The qualitative assessment shown in Table 2 explains the stages for the pedestrian facilities' condition, starting from the first condition describing pedestrians without facilities and improved gradually until the fifth condition which represents the most convenient facilities for pedestrians. Visual photos from several pedestrian locations have been used for references in the pilot project. Consequently, these photos did not represent the actual conditions (as shown in Table 2). As a result, most of the respondents were confused while answering the question sheets. After evaluation the pilot project, the visual photos were replaced by visual illustrations to exemplify the parameter of qualitative measurement. The visual illustrations used the same concept for each condition. Such as color differentiation for added/elimination value for each stage (Fig. 1).

3 Results and Discussion

In the beginning using a visual photo caused a disorientation for the respondents. Then, we conducted the preliminary survey using visual illustrations in Sudirman Integrated Railway Area which has a MRT station (Dukuh Atas) and commuter line station (Sudirman). The survey was distributed using a Google Form since June 10–July 10, 2022, and the characteristic of the respondent is shown in Fig. 2. As

Table 2 The stages for the pedestrian facilities condition

Second2nd Parameter Convenience				
 1st	 2nd	 3rd	 4th	 5th
Unfacilitated pedestrian	Most pedestrian path is broken	There are some big holes	Some tiles are missing, Uneven level	Convenience pedestrian path
3rd Parameter Safety				
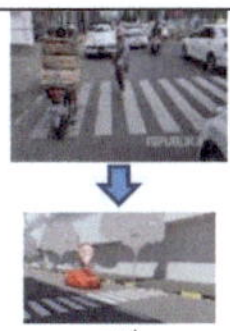 1st	 2nd	 3rd	 4th	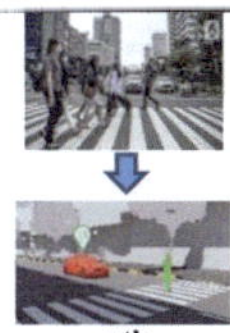 5th
• High-speed vehicles • Impolite riders	• The max speed of vehicles is 40 km/hr	• The max speed of vehicles is 40 km/hr	• The speed of vehicles between 20 to 40 km/hr	• The speed of vehicles is below 20 km/hr

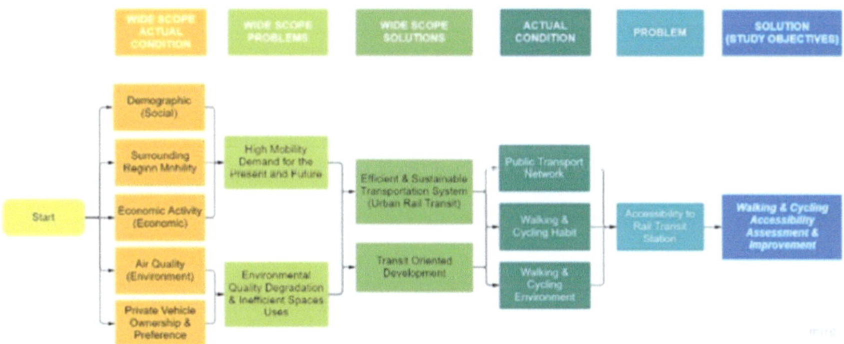

Fig. 1 Research framework

can be seen in this figure, men contributed more than women in responding to the survey. The participants are dominated by people on productive ages who are between 23–55 years old (63.64%) which is in line with the land use in Sudirman zone as central business district. Walking ability is linked with the daily walking habits of the respondents. A quarter of respondents usually walk for 16–20 min in one day. A higher percentage of the respondents are respondents who have a daily habit of walking for 10–15 min and 26–35 min. In Table 3, one can observe that men were more preferably walk longer than women. In total the respondents who responded

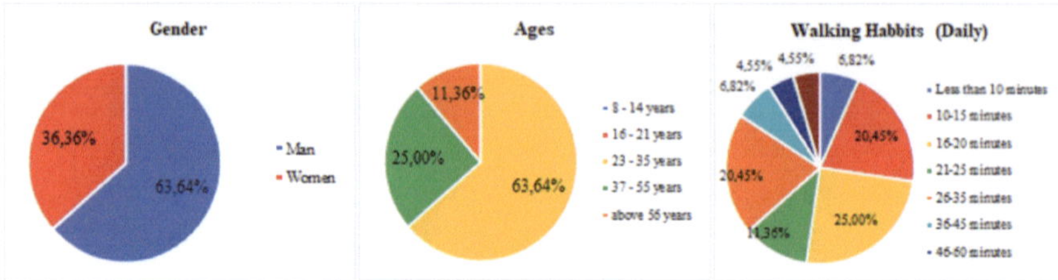

Fig. 2 Characteristics of respondents

to the survey were people who liked/prefer to walk. The summary of the qualitative assessment based on respondents' assumptions can be seen in Table 4 and Fig. 3. Most of the respondents accepted walking in inappropriate facilities which is in line with the walking habits of the respondents. The acceptance rate for walking, in the aspect of easiness, in male participants is lower than that in female participants because most of the male participants decided to walk even though there are barriers on the pedestrian path. Furthermore, male and female participants will continue to walk despite there being many holes in the pedestrian path. The result of the visually illustrated questionnaire is the same as the authors' expectation that it will give a better and good response.

Table 3 Walking habits of respondents

Walking habits (Daily)	Less than 10 min	10–15 min	16–20 min	21–25 min
Men	3.57%	10.71%	21.43%	14.29%
Women	12.50%	37.50%	31.25%	6.25%
Walking habits (Daily)	26–35 min	36–45 min	46–60 min	Above 60 min
Men	32.14%	10.71%	3.57%	3.57%
Women	0.00%	0.00%	6.25%	6.25%

Table 4 The respondent's perspective about the pedestrian facilities

Indicators	The stages of conditions for the pedestrian facilities				
	1st condition	2nd condition	3rd condition	4th Condition	5th condition
1st Parameter Easiness (Barrier)					
Man	21.43%	28.57%	14.29%	25.00%	10.71%
Woman	25.00%	25.00%	12.50%	6.25%	31.25%
2nd Parameter Convenience (Quality of Pedestrian)					
Man	14.29%	32.14%	10.71%	28.57%	14.29%
Woman	25.00%	33.33%	8.33%	8.33%	25.00%
3rd Parameter Safety (Crossing)					
Man	11.76%	29.41%	17.65%	26.47%	14.71%
Woman	33.33%	33.33%	11.11%	0.00%	22.22%

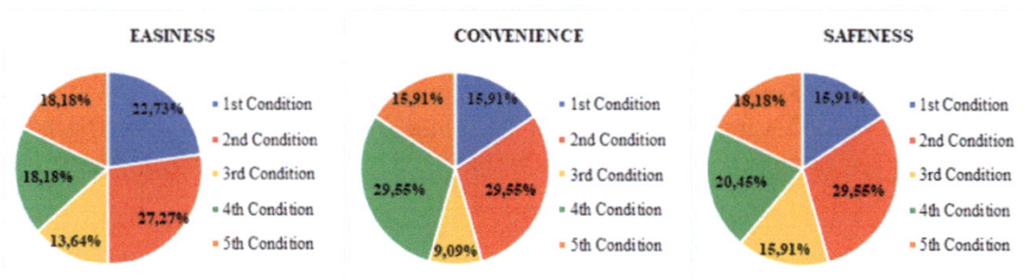

Fig. 3 The respondent's perspective about the pedestrian facilities

4 Conclusion

There are numerous methodologies to assess walkability and bikeability. In this research we combined quantitative and qualitative assessment. In measuring walkability, the author combined the perspective of the passenger trains (LOA) with the pedestrian facilities and (PLOS) while for bikeability assessment using BLOS. In Greater Jakarta, there are 126 rail stations; it's impossible to assess all the station for research. Therefore, we conduct elimination staging process. As a result, 27 stations were chosen as the research study locations.

The qualitative assessment requires the respondents to elaborate on the pedestrian conditions in the same manner and same perspectives. In the beginning, using a visual photo caused disorientation for the respondents. Therefore, in this study, visual illustrations were made to make the qualitative assessments easier to understand for respondents. In the preliminary survey in the Sudirman Area, the result shows better responses from the respondents and surveyor.

Acknowledgements The researchers would like to highly appreciate P2MI ITB for funding this research and the collaboration between ITB and Telkom University.

References

1. E. Vazquez, B. Rodrigues, M.K. Najjar, A. Haddad, A.W. Hammad, A framework to locate new medium and high-capacity stations for public transportation modes of transport in Rio de Janeiro. Clean. Eng. Technol. **10**(2022). https://doi.org/10.1016/j.clet.2022.100537
2. C. Garau, G. Desogus, B. Barabino, M. Coni, Accessibility and public transport mobility for a smart (er) Island: evidence from sardinia (Italy). Sustain. Cities Soc. **87**(August), 104145 (2022). https://doi.org/10.1016/j.scs.2022.104145
3. S. Nawaz, Y. Ali, Analyzing the influence of social, cultural, behavioral traits on cycling and walking in Pakistan. Transp. Res. Interdiscip. Perspect. **7**(2020). https://doi.org/10.1016/j.trip.2020.100182
4. J. Gao, M. Helbich, M. Dijst, C.B.M. Kamphuis, Socioeconomic and demographic differences in walking and cycling in the Netherlands: How do these translate into differences in health benefits? J. Transp. Health **6**, 358–365 (2017). https://doi.org/10.1016/j.jth.2017.06.001

5. H. Marquart, J. Schicketanz, Experiences of safe and healthy walking and cycling in urban areas: The benefits of mobile methods for citizen-adapted urban planning. Transp. Res. Procedia **60**, 290–297 (2022). https://doi.org/10.1016/j.trpro.2021.12.038
6. B. Jardim, M. de Castro Neto, Walkability indicators in the aftermath of the COVID-19 pandemic: a systematic review. Sustainability **14**(17), 10933 (2022). https://doi.org/10.3390/su141710933
7. W. Hadi, Y. Chrisnawati, H.N. Ikhsan, Public transportation accessibility: towards sustainable transit oriented development (Case study: Depok Baru Station-Jakarta, Indonesia. IOP Conference Series: Earth and Environmental Science **202**(1), 012012 (2018)
8. Irmawandari, K. D. Handayeni, Kajian Aksesibilitas Stasiun dengan Moda Berjalan Kaki di Kota Surabaya. Jurnal Teknik ITS, **8**(2), E51 (2020)
9. T. Althoff, R. Sosič, J.L. Hicks, A.C. King, S.L. Delp, J. Leskovec, Large-scale physical activity data reveal worldwide activity inequality. Nature **547**(7663), 336–339 (2017)
10. R. Bona Frazila, F. Zukhruf, J. Jihanny, K. Nawassa, E. Zusman, R. Nakano, S. Budi Nugroho, Developing walkability index from walker and non-walker perception. in *Proceedings of the Eastern Asia Society for Transportation Studies*, **11**, 2017 (n.d.)
11. S.S. Wibowo, P.S. Olszewski, Using equivalent walking distance to assess pedestrian accessibility to transit stations in Singapore. Transp. Res. Rec. **1927**(1), 38–45 (2005)
12. R.A. Jayanti, T.B. Joewono, R. Muhamad, Aksesibilitas Stasiun KRL Berdasarkan Persepsi Wanita. J. Transp. **21**(1), 63–72 (2021)
13. M. Özyavuz, T. Özcan Aktan, Criteria of walkability: as a sense of urban experiment tekirdağ atatürk ormani tabiat parkinin sicaklik dağilimi ve çevresel etkisi view project criteria of walkability: as a sense of urban experiment. J. Balk. Sci. Technol., **2022**(1), 73–80. (2022). https://doi.org/10.55848/jbst.2022

Study of Ship Handling in Wheel and Steering Commands for Simulator and Onboard Ship Training of Nautical Students

Rasyidah Mohd Sani, Mohammed Ismail Russtam Suhrab, and Nik Muhamad Afiz Harom Junoh

1 Introduction

Proficiency in ship handling is a crucial competency for nautical students, demanding a deep understanding of how ships respond to diverse environmental conditions and the adeptness to efficiently control their movements [1]. To navigate a ship in confined spaces, crowded harbors, and challenging docking scenarios, ship handling skills are indispensable for marine operations [2, 3]. Despite the challenges associated with this skill, ship handlers take pride in their ability to maintain control while the ship is in motion [4].

The occurrence of ship steering control failures is not uncommon, particularly in narrow channels or amid heavy maritime traffic, as Murphy's law states, and tends to manifest at the most opportune moments [5]. Unfortunately, human errors are frequently blamed for these occurrences [6, 7]. The ship's nerve center for movement control is a navigation bridge equipped with an array of steering and control mechanisms, including autopilots, dynamic positioning, and tracking units (DP/DT), steering wheels, tillers, joysticks, pitch controllers, bow thruster controls, and joysticks [8]. Larger vessels boast multiple control stations distributed across bridges. Regardless of sophisticated control systems, incidents such as the tanker Halit Bey Grounding at St. In 2014, the Lawrence River served as a harsh reminder of the difficulties associated with ship handling. The grounding of the tanker was

R. M. Sani · M. I. R. Suhrab (✉) · N. M. A. H. Junoh
Faculty of Maritime Studies, Universiti Malaysia Terengganu, Kuala Terengganu, Malaysia
e-mail: m.ismail@umt.edu.my

R. M. Sani
e-mail: s55177@ocean.umt.edu.my

N. M. A. H. Junoh
e-mail: nm.afiz@umt.edu.my

A. Fitriadhy et al. (eds.), *Advancing Interdisciplinary Knowledge and Innovation*, SpringerBriefs in Applied Sciences and Technology,
https://doi.org/10.1007/978-3-032-01309-5_7

attributed to a steering control malfunction, underscoring the critical importance of a comprehensive understanding of various control systems and their nuances [9].

In restricted areas, such as rivers, canals, shipping channels, and ports, the Officer on Watch (OOW) must act swiftly to retake control of the ship when conventional steering systems fail. In these situations, there is no room for delayed reactions, and diagnosing the issue can be more challenging because of the involvement of many technological components [10]. As a protective measure, maritime law mandates the inclusion of an auxiliary control system, commonly known as an emergency backup system [11]. The role of human error in ship handling cannot be overstated, as it is responsible for 80% of marine accidents, often occurring during maneuvers intended to avoid collisions [12, 13]. To ensure precision and uniformity in the issuance of rudder angle commands, the International Maritime Organization (IMO) has established standardized wheel directions that are widely used in industry [14].

The COVID-19 pandemic has disrupted conventional pedagogical approaches for nautical students [15]. Institutions such as the University Malaysia Terengganu (UMT) have shifted toward online theory-based learning, resulting in limited opportunities for practical firsthand training in ship simulation studies [16]. Furthermore, the practical skills taught in these online-based settings can be easily forgotten without practice [17]. In this context, this study aimed to assess the knowledge and skills of nautical science students regarding ship handling, especially regarding wheel and steering commands for simulator and onboard ship training. Additionally, it seeks to compare the performance, knowledge, and ship-handling abilities of nautical science students before and after simulator and onboard ship training, offering recommendations for enhancing ship handling in simulator and onboard ship training.

2 Research Methodology

The research starts out by heavily relying on case studies, a method that is efficient in answering difficult "how" and "why" issues. Case studies make it easier to gather comprehensive qualitative data using a variety of sources including records, observations, and interviews. This integrative method ensures a comprehensive comprehension of the topic while improving the reliability and validity of the study. The research methodology consisted of two fundamental components: data collection and data analysis.

A qualitative approach was used to achieve the research objective. Six principal areas were covered: how ship bridge simulators can help nautical students learn more, how to evaluate students' knowledge of wheel commands and attitude toward steering before and after simulator and ship handling training, and how to obtain feedback on how to make training more effective. To access knowledge and skills, a Likert scale questionnaire was utilized. Research papers, books, notes, official websites, and certifications were the primary materials. Organizing the presenting data, charts, Microsoft Excel and Word, Google forms for administering questionnaires, and analytical hierarchy process (AHP) software were employed. Five candidates participated in a pilot study to identify any problems with the questionnaire. The data analysis was performed as follows.

The data analysis procedure included three analytical approaches: Arithmetic mean (AM), analytical hierarchy process (AHP), and hierarchical structural model of research. The sum of the numerical values of every observation divided by the total number of comments is the arithmetic mean of a collection of observed data. In a prescribed way, if we have a data set of values, $a_1, a_2 \ldots\ldots a_n$, then the arithmetic mean is defined by Eq. (1):

$$\mathrm{A} = \frac{1}{n}\sum_{i=1}^{n} a_i \text{ also as same as } \mathrm{A} = \frac{\sum x_1 + x_2 + x_3 + x_4 + x_5 + x_n}{n}. \quad (1)$$

where the average of the criteria intensity = A; the number of respondents = n; number of criteria = a_i;; the intensity of scale (1–9) denoted as x; n represented as respondent = x_n.. The next approach is the analytical hierarchy process (AHP), which is an effective approach for evaluating assessment criteria and alternative possibilities for executing decisions. To assess the relative weights of the criteria and alternatives, pairwise comparisons were used, which resulted in extensive ranking. Furthermore, it offers a well-organized framework for decision-making. Finally, the evaluation of knowledge and skills in ship handling is illustrated by the hierarchical structural model of research (Fig. 1), which uses a hierarchy of criteria and alternatives. The criteria comprise skills and knowledge, and knowledge alternatives include navigation, communication, and ship stability. Furthermore, teamwork, efficiency, and decision-making are examples of skill alternatives. A flowchart of the overall research methodology is shown in Fig. 2.

3 Results and Discussion

Forty nautical students from Batches 13 to 15 of the University Malaysia Terengganu participated in this study. Twenty respondents went through a simulator, and 20 more were onboard training on ship handling in wheel and steering commands.

Based on Table 1, skills (0.6630) were in first place followed by knowledge (0.3370) in second place before simulator training. For onboard training, skills (0.6420) are also in the first place, followed by knowledge (0.3580). This shows the importance of ship handling skills in wheel and steering commands. In addition, skills boosted driving confidence. It enhances communication, time management, negotiation abilities, and the ability to expand one's professional networks. Developing their skills is something that they can do through learning. Identification, development, and fostering of intrinsic abilities in the intended sector are all made possible through skill training.

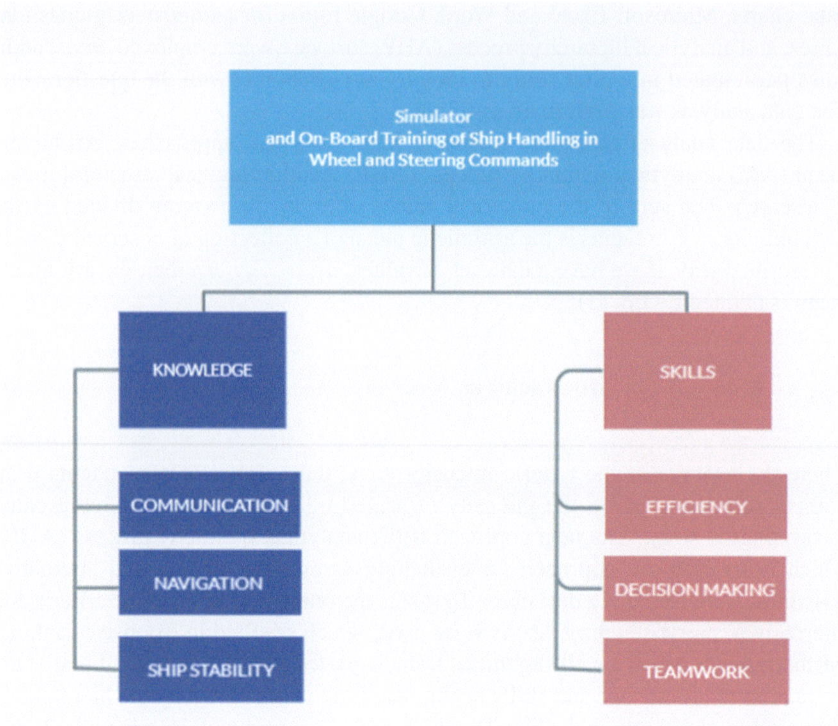

Fig. 1 Hierarchy structural model

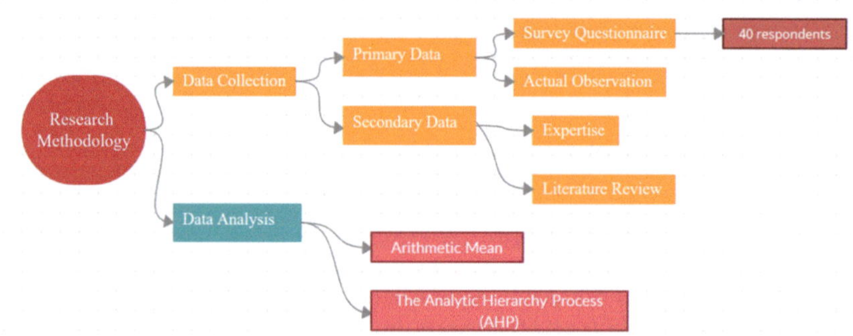

Fig. 2 Research method

Table 1 The comparative analysis (pair wise comparison) of the AHP methods before training

Training	Element	Sub-element	Local weight	Global weight
Simulator	Knowledge (0.337)	Communication	0.413	0.1392
		Navigation	0.42	0.1415
		Ship stability	0.167	0.0563
	Skills (0.663)	Efficiency	0.258	0.1711
		Decision making	0.325	0.2155
		Teamwork	0.417	0.2765
Onboard	Knowledge (0.358)	Communication	0.64	0.2291
		Navigation	0.171	0.0612
		Ship stability	0.189	0.0677
	Skills (0.642)	Efficiency	0.238	0.1528
		Decision making	0.252	0.1618
		Teamwork	0.509	0.3268

Table 2 shows skills (0.5890) in the first place, followed by knowledge (0.411) in the second place after simulator training. For onboard training, skills (0.6020) were also in the first place, followed by knowledge (0.3980). This shows the importance of ship handling skills in wheel and steering commands. Additionally, without knowledge, they do not know how to steer and depend only on skills. If they do not know about the maritime phrase Convention on the International Regulations for Preventing Collisions at Sea, 1972 (COLREGs), they will face problems.

Table 2 The comparative analysis (pair wise comparison) of the AHP methods after training

Training	Element	Sub-element	Local weight	Global weight
Simulator	Knowledge (0.411)	Communication	0.670	0.2754
		Navigation	0.199	0.0818
		Ship stability	0.131	0.0538
	Skills (0.589)	Efficiency	0.477	0.2810
		Decision making	0.177	0.1043
		Teamwork	0.347	0.2044
Onboard	Knowledge (0.398)	Communication	0.635	0.2527
		Navigation	0.216	0.0860
		Ship stability	0.149	0.0593
	Skills (0.602)	Efficiency	0.347	0.2089
		Decision making	0.281	0.1692
		Teamwork	0.373	0.2245

Table 3 Normalized value in percentage of main element before and after training

Training	Main element	Percentage (%)	
		Before	After
Simulator	Knowledge	48	80
	Skills	57	82.6
Onboard	Knowledge	55.4	81.3
	Skills	63.1	85

Table 3 shows an element of knowledge (48%) and skills (57%) before simulator training. The respondent was weak in ship handling in wheel and steering commands. Prior training on board, knowledge (55.4%) and Skills (63.1%). The respondents had minimal experience and practice in simulators and onboard ships. They also tended to forget what they learned in class. The two elements of knowledge and skills increased significantly after simulator and onboard training. After simulator training, element knowledge (80%) and skills (82.6%) were acquired, while after onboard training, element knowledge (81.3%) and skills (85%) were acquired. This shows the vital knowledge and skills of ship handling in terms of wheel and steering commands.

4 Conclusion

In conclusion, both simulator and onboard training are cost-effective alternatives for training competent deck cadets who can provide significant outputs in complex and unexpected situations. This study employed the arithmetic mean to analyze data from 40 respondents evenly distributed between the simulator and onboard training groups. The process of assigning weights to evaluation criteria was facilitated using the analytical hierarchy process (AHP). The analytical hierarchy process (AHP) further aided in assigning weights to the evaluation criteria. This research encompasses two primary elements, each with three sub-elements or alternatives. AHP allows for a hierarchical structure, provides a focused assessment of specific criteria and sub-criteria, and enables informed decisions and ranking based on evaluations. The study findings demonstrate that, after receiving simulator and onboard training, students enhance their skills and capabilities in ship managing strategies, further promoting heightened situational awareness and self-assurance. However, it is important to manage the constraints on practical time during simulator sessions and recommend increasing the training time to ensure that students acquire comprehensive knowledge and skills.

Despite the study's insight, it is important to acknowledge some limitations of this research, including the comparatively small sample size because of the limited number of nautical science students at the University of Malaysia Terengganu. Furthermore, the findings of some analyses, such as multiple regression analysis, may take time to obtain. In the future, it will be necessary to create a more accurate evaluation model. Improved solutions should be developed and coordinated with the results of more thorough and in-depth studies. To guarantee that students studying nautical science are incredibly well prepared for their future careers as sailors, ship handling instruction must be continuously improved.

Acknowledgements We acknowledge the Nautical Science students from Batches 13 to 15 who participated in this research and provided valuable information for this research. We are thankful to the Faculty of Maritime Studies, Research and Field Service Centers of University Malaysia Terengganu for allowing us to conduct this study.

References

1. C. Sellberg, O. Lindmark, H. Rystedt, Learning to navigate: the centrality of instructions and assessments for developing students' professional competencies in simulator-based training. WMU J. Marit. Aff. **17**, (2018). https://doi.org/10.1007/s13437-018-0139-2
2. R.G. Wright, Unmanned and autonomous ships: an overview of MASS. Unmanned and Autonomous Ships. (2020)
3. A. Carse, The Feel of 13,000 Containers: how pilots learn to navigate changing logistical environments. Ethnos. **88**, (2023), https://doi.org/10.1080/00141844.2019.1697337
4. M.H. Sliem, E.M. Fayyad, A.M. Abdullah, N.A. Younan, N. Al-Qahtani, F.F. Nabhan, A. Ramesh, N. Laycock, M.P. Ryan, M. Maqbool, D. Arora, Monitoring of under deposit corrosion for the oil and gas industry: A review. J. Pet. Sci. Eng. J PETROL SCI ENG, **206** (2021)
5. M.M.H. Imran, A.F. Ayob, S. Jamaludin, Applications of artificial intelligence in ship berthing: A review, Indian J. Geo-Mar. Sci, **50**(11), (2021)
6. S. Fan, J. Zhang, E. Blanco-Davis, Z. Yang, X. Yan, Maritime accident prevention strategy formulation from a human factor perspective using Bayesian Networks and TOPSIS. Ocean Eng. **210**, (2020). https://doi.org/10.1016/j.oceaneng.2020.107544
7. K. Wróbel, Searching for the origins of the myth: 80% human error impact on maritime safety, Reliab. Eng. Syst. Saf. **216**, (2021)
8. L. Liu, X. Guo, W. Liu, C. Lee, Recent progress in the energy harvesting technology—from self-powered sensors to self-sustained iot, and new applications, nanomaterials, **11**, (2021)
9. B. Göksu, O. Yüksel, C. Şakar, Risk assessment of the Ship steering gear failures using fuzzy-Bayesian networks. Ocean Eng. **274**, (2023). https://doi.org/10.1016/j.oceaneng.2023.114064
10. A. Gerson, Stranding of the mega-ship ever given in the suez canal: causes, consequences, and lessons to be learned. (2023)
11. K. Hiekata, S. Wanaka, T. Mitsuyuki, R. Ueno, R. Wada, B. Moser, Systems analysis for deployment of internet of things (IoT) in the maritime industry. J Mar Sci Technol J Mar Sci Tech-Japan. **26**, (2021). https://doi.org/10.1007/s00773-020-00750-5
12. W.J. Son, J.S. Lee, H.T. Lee, I.S. Cho, An investigation of the ship safety distance for bridges acrosswaterways based on traffic distribution. J. Mar. Sci. Eng. **8**, (2020). https://doi.org/10.3390/JMSE8050331
13. S. Cope, E. Hines, R. Bland, J.D. Davis, B. Tougher, V. Zetterlind, Application of a new shore-based vessel traffic monitoring system within san francisco bay. Front Mar Sci. **7**, (2020). https://doi.org/10.3389/fmars.2020.00086
14. A.S. Kamis, A.F.A. Fuad, A.Q. Anwar, Systematic literature review on wheel-over point techniques for efficient ship manoeuvre. Trans. Marit. Sci. **11**, (2022). https://doi.org/10.7225/toms.v11.n02.006
15. Demirel, E.: Impact of Covid 19 pandemic on maritime education and training. In: International Scientific Conference SEA-CONF (2021)

16. K. Johansen, K.: Challenges regarding digital distance learning of operationally- oriented professions, due to Covid-19 pandemic. Int. J. Educ. Res. **4**, (2023). https://doi.org/10.1016/j.ijedro.2023.100225
17. G.S. Ochavillo, A paradigm shift of learning in maritime education amidst COVID-19 Pandemic. Int. J. High. Educ. **9**, (2020). https://doi.org/10.5430/ijhe.v9n6p164

Enhancing Safety: Investing the Impact of Support Services on Accident Reduction in Chittagong Port and Outer Anchorage of Bangladesh

Tamim Ahmed, Mohammed Ismail Russtam Suhrab, Ahmad Faizal Ahmad Fuad, Noor Apandi Osnin, and Kasypi Mokhtar

1 Introduction

The marine sector is the largest in the world's transportation system, and its growth is continually increasing. The overall volume of seaborne trade exceeded 11 billion tons in 2018, and the development of international trade and the global economy depended on seaports [1]. Maritime transportation continues to rule out other forms of transportation for international trade. Globalization and rising demand for maritime transport services from emerging economics are credited with this success [2]. This comprehensive success would not have been possible without considering capacity improvement, seaport operation safety, and operational process efficiency.

Chittagong port operations support cargo handling operations for more than 80% of the world's volumetric merchandise trade moved by the sea. Maritime shipping is a significant high-risk business. Reducing the risks posed by this industry and enhancing maritime safety are significant concerns, numerous factors can improve maritime safety and reduce the likelihood of accidents. Most of these factors relate to enhancing operational risk management systems, leveraging innovative machinery and equipment, and applying appropriate simulation training to reduce unintentional

T. Ahmed · M. I. R. Suhrab (✉) · A. F. A. Fuad · N. A. Osnin · K. Mokhtar
Faculty of Maritime Studies, University Malaysia Terengganu, Kuala Terengganu, Malaysia
e-mail: m.ismail@umt.edu.my

T. Ahmed
e-mail: capt.tamim73@gmail.com

A. F. A. Fuad
e-mail: afaf@umt.edu.my

N. A. Osnin
e-mail: apandi@umt.edu.my

K. Mokhtar
e-mail: kasypi@umt.edu.my

A. Fitriadhy et al. (eds.), *Advancing Interdisciplinary Knowledge and Innovation*,
SpringerBriefs in Applied Sciences and Technology,
https://doi.org/10.1007/978-3-032-01309-5_8

demurrage. One of these characteristics, humans, might be thought of as the fulcrum of maritime safety [3].

The Chittagong port is the primary seaport in Bangladesh. It provides a significant gateway to a country's trade with the outside world. According to an overview of Chittagong port (CP) (2013), during 2011–2012, Chittagong port handled over 41 million metric tons of cargo, including 1.34 million TEUs, which is approximately 92% of the total maritime trade in Bangladesh. However, Chittagong port has the highest accident rate among all seaports in Bangladesh. Khaled et al. (4) pointed out that the Bangladesh waterway accident database for the period 1981–2013 shows 57.96% of accidents were due to collision and grounding, 76.92% at Chittagong port. The maximum number of accidents occurred because of human error. They also mentioned that most accidents from 1981 to 2013 at Chittagong port occurred because of human error, and the rate was 75.36% among all casualties [4].

As shown in Table 1, human error causes approximately 62% of accidents in Bangladesh, and environmental factors cause 33% of accidents. The percentage of collisions in Chittagong port was 33% of collisions recorded in Bangladesh since 1981, and navigational accidents caused were 80% [5].

All of these factors affect the performance of employees in the Chittagong port. Al-Shammari and Jin-Seok [6] pointed out that accidents in the Chittagong anchorage result from maneuvering vessels failing to consider the variability and strength of the tide and currents, leading to contact between anchored and embarking vessels. It was observed that most employees quit Chittagong port to look for jobs with minimal accidents. Based on the accidents in Chittagong port, Aydin [7] observed that many ships collided owing to poor communication and lost many lives. Apart from these problems, the researcher noted that Chittagong port does not provide maritime support, operational risk management, simulation training, and assessment, which should help employees reduce accidents. This serves as a gap to be addressed in this study. The authors conducted a study to assess the relationship between operational risk management, maritime support services, and simulation training and assessment within accident reduction in Chittagong port and the outer anchorage of Bangladesh.

Table 1 Percentage of accidents in chittagong port and outer anchorage, Bangladesh

Factors	In all accidents in Bangladesh (%)	In all accidents at CP (%)	In all collisions at CP (%)
Human error	62	75.36	57.57, 87.87
Environment error	33	8.7	33.33
Technical error	5	15.94	9.09

2 Methodology

A descriptive survey was conducted. The sample consisted of 100 employees from Chittagong port, Bangladesh. Therefore, a simple random sampling technique was used. Three research hypotheses were proposed to guide this study. The data for the hypotheses were analyzed using Pearson's moment correlation coefficient.

The study used three variables: maritime support services, operational risk management, and simulation training and assessment. The instrument was divided into four parts. The first part contained the respondents' information, such as gender, age, experience, and qualifications. The second part, maritime support, had six items. A third of the instruments were operational risk management, with six items; the last was an employee performance instrument, with eight items, while simulation training and assessment had eight items. Based on this, some questions were adapted from researchers, such as maritime support services [8], operational risk management, and training for simulation and assessment [9, 10]. Therefore, the total number of items used for exercise was 20 for all three variables. Descriptive statistics were used to analyze all respondents' demographic data. Three hypotheses were formulated for this study. All hypotheses were analyzed using the Pearson moment correlation coefficient.

3 Results and Discussions

The respondents' characteristics used in this study are listed in Table 2. Some respondents' characteristics included gender, age, work experience, qualifications, and designation. The total number of participants in the final study was 100. According to the analysis, the number of male respondents participating in the study was 70 (70.0%), while that for female was 30 (30.0%). This shows that there were more male respondents than female respondents in Chittagong port, Bangladesh. Concerning the age group, the analysis showed that respondents' age group for 20–30 years had a frequency of 16 (16.0%), followed by the age group for 31–40 years with a frequency of 32 (32.0%), the group for 41–50 years had a frequency of 29 (29.0%), and the age group for 51–60 years had a frequency of 23 (23.0%). The results indicated that the age group was in the range 31–40 years because it shows a high frequency in the distribution.

The authors applied three hypotheses to fulfill the study's objectives. The first hypothesis was that there would be a significant relationship between operational risk management and accident reduction in Chittagong port and outer anchorage in Bangladesh. A Pearson product-moment correlation was run to assess the relationship between operational risk management and accident reduction in Chittagong port and outer anchorage in Bangladesh. The correlation coefficient size is from 0.91 to 1.00 or −0.91 to −1.00, indicating a very strong correlation; 0.71 to 0.91 or −0.71 to −90 indicates a strong relationship; 0.51 to 0.70 or −0.51 to −0.70 is average/medium;

Table 2 Demographic profile of the interviewed person during the study

Demographic	Categories	Frequency	Percentage
Gender	Male	70	70.0
	Female	30	30.0
Age	20–30 yrs	16	16.0
	31–40 yrs	32	32.0
	41–50 yrs	29	29.0
	51–60 yrs	23	23.0
Marital status	Married	75	75.0
	Single	25	25.0
	Diploma	10	10.0
Qualifications	Bachelor	35	35.0
	Masters	44	44.0
	Ph.D	11	11.0
Years of working experience	1–10	34	34.0
	11–20 yrs	19	19.0
	21–30 yrs	34	34.0
	31–40 yrs	13	13.0
Designation	Executive	23	23.0
	Technical supervisor	31	31.0
	Senior officer	24	24.0
	Junior staff	22	22.0

0.31 to 0.50 or −0.31 to −50 is weak; 0.01 to 0.30 or −0.01 to −30 is very weak while 00 indicates no correlation. The correlation between the two variables was analyzed, with $r = 0.858$, $n = 100$, and $P < 0.005$, indicating a strong correlation between operational risk management and accident reduction. In this positive correlation, it was observed that the higher the value of operational risk management, the higher the value of accident reduction (Table 3).

The second hypothesis was that there would be a significant relationship between maritime support services and accident reduction in the Chittagong port. The correlation between the two variables was analyzed, with $r = 0.611$, $n = 100$, and $P < 0.005$, indicating a moderate correlation between maritime support and accident reduction. In this positive correlation, it is observed that the correlation coefficient indicates a relationship hence confirming the second hypothesis, i.e., the higher the value of maritime support, the higher the value of accident reduction (Table 3).

The third hypothesis was the relationship between simulation training and assessment and accident reduction in Chittagong port and outer anchorage in Bangladesh.

Table 3 Correlation between tested three hypotheses

Parameters		Accident reduction	Hypothesis
Operational risk management	Pearson correlation	0.858**	
	Sig. (2-tailed)	0.000	H1 supported
	N	100	
Maritime support	Pearson Correlation	0.611**	
	Sig. (2-tailed)	0.000	H2 supported
	N	100	
Simulation training and assessment	Pearson correlation	0. 810**	
	Sig. (2-tailed)	0.000	H3 supported
	N	100	

** Correlation is significant at the 0.01 level (2-tailed)

The correlation between the two variables was analyzed, where r = 0.810, n = 100, and P <0.005, indicating a strong correlation between simulation training, assessment, and accident reduction. In this positive correlation, it was observed that the higher the simulation training and assessment value, the higher the accident reduction value (Table 3). This study explores the relationship between support services and accident reduction in Chittagong port and outer anchorage in Bangladesh. The study's conclusion was similar to the survey by Mokhtari et al. [11] where top management must consider operational risk management. The findings of this study confirmed a significant relationship between active risk management and accident reduction (r = 0.858).

Similarly, a significant relationship was found between maritime support and accident reduction (r = 0.611). Therefore, this finding corroborates the findings of other studies [12–15] whose support is crucial in the maritime industry because it increases the productivity of employees and reduces envisaged accidents.

4 Conclusions and Recommendations

The research was conducted at the Chittagong port in Bangladesh, and the conceptual framework was developed based on existing leadership theories and relevant materials. The results of the study indicated that employees' experience working at the Chittagong port reflected their performance had been hindered. However, the study found that operational risk management, maritime support services, and simulation training and assessment had a significant positive relationship with accident reduction. The findings of this study demonstrate that employees' performance can be improved through support services.

As the world undergoes rapid changes, the maritime sector is also impacted. To keep pace with these changes, it is necessary to provide additional support services to employees to allow them to carry out their duties effectively and reduce accidents in the organization. Technological advancements in the maritime sector require employees to have the necessary knowledge and skills to use the latest technologies, which can be acquired through support services and training mechanisms. The implementation of these measures will enhance employee effectiveness and prevent accidents from occurring.

References

1. F. Moreno, J. Gonzalez, J. Muro, Relationship between human factors and the safe performance of vessel traffic service operators: a systematic qualitative-based review in maritime safety. Saf. Sci. **155**(2), 125–140 (2022). https://doi.org/10.1016/j.ssci.2022.105892
2. W. Lin, H. Cheng, Improving maritime safety through enhancing marine process management: The application of balanced scorecard. Man. Dec. **59**(3), 604–615 (2022). https://doi.org/10.1108/MD-08-2019-1044
3. O. Melnyk, S. Onyshchenko, N. Mavlova, Integrated ship cybersecurity management as a part of maritime safety and security system. Int. J. Com. Sci. Nat. Sci. **22**(03), 135–140 (2022). https://doi.org/10.22937/IJCSNS.2022.22.3.18
4. L. Chuah, K. Mokhtar, A. Bakar, M. Othman, Marine environment and maritime safety assessment using Port State Control database. Chem. **304**(2), 135–150 (2022). https://doi.org/10.1016/j.chemosphere.2022.135245
5. B. Mindong, T. Yiping, Y. Yixuan, Application of a hydroxyl-radical-based disinfection system for ballast water. Chem. **208**(1), 541–549 (2018). https://doi.org/10.1016/j.chemosphere.2018.06.010
6. L. Chuah, N. Alam, J. Daniel, Investigating the crude oil biodegradation performance in the bioreactor by using a consortium of symbiotic bacteria. Chem. **337**(2), 95–105 (2023). https://doi.org/10.1016/j.chemosphere.2023.139293
7. N. Al-Shammari, J. Oh, Effects of human error on marine safety: A case study. J. Eng. Res. App. **8**(1), 11–19(2018). https://doi.org/10.9790/9622-0811021119
8. C. Mangga, P. Tibo-oc, R. Montaño, Impact of engine room simulator as a tool for training and assessing BSMARE students' performance in engine watch keeping. Пед. **93**(6), 88–100 (2021). https://doi.org/10.53656/ped21-6s.07eng
9. M. Abdoli, M. Falahnezhad, S. Behboudian, Multivariate econometric approach for solid waste generation modeling: a case study of Mashhad, Iran. Env. Eng. Sci. **28**(9), 627–633 (20122). https://doi.org/10.1089/ees.2010.0234
10. C. Sellberg, Stimulators in bridge operations training and assessment: a systematic review and qualitative synthesis. WMU. J. Mar. Aff. **16**(1), 247–263 (2017). https://doi.org/10.1007/s13437-016-0114-8
11. K. Mokhtari, J. Ren, C. Roberts, Decision support framework for risk management on seaports and terminals using fuzzy set theory and evidential reasoning approach. Exp. J. Syst. App. **39**(5), 50–87 (2012). https://doi.org/10.1016/j.eswa.2011.11.030
12. J. Sylvia, N. Anupam, Doing well by talking good: a topic modelling-assisted discourse study of corporate social responsibility. App. Ling. **39**(3), 373–399 (2018). https://doi.org/10.1093/applin/amw014
13. T. Kim, A. Sharma, M. Bustgaard, The continuum of simulator-based maritime training and education. WMU J. Mar. Aff. **3**(1), 135–150 (2021). https://doi.org/10.1007/s13437-021-00242-2
14. S. Pileggi, M. Indorf, A. Nagi, CoRiMaS—an ontological approach to cooperative risk management in seaports. Sust. **12**(11), 47–67 (2021). https://doi.org/10.3390/su12114767

15. C. Wiig, C. Sellberg, M. Solberg, Reviewing simulator-based training and assessment in maritime education: a topic modeling approach for tracing conceptual developments. WMU J. Mar. Aff. **22**(2), 143–164 (2023). https://doi.org/10.1007/s13437-023-00307-4

A Study on the Establishment of Vessel Traffic Services at Penang Port

Ahmad Faizal Ahmad Fuad, Nur Iman Insyirah Zamri, Amir Syawal Kamis, and Mohd Saiful Izwaan Saadon

1 Introduction

Managing marine traffic is crucial for reducing accidents. Vessel traffic services oversee ship traffic, advising on routes and preventing collisions [1, 2]. They are essential for safety at sea, navigation efficiency, and environmental protection. Their role has grown significantly in congested ports and waterways, enhancing safety and efficiency while protecting the environment [3, 4]. The competent authority has implemented VTS to improve traffic safety, efficiency, and environmental protection in sensitive marine areas and high-traffic zones [5].

The Penang Port is located in the north of the west coast of the Peninsular Malaysia and has three terminals, namely the North Butterworth Container Terminal, Bulk Cargo Terminal, and the Swettenham Pier Cruise Terminal. In terms of cargo volume, the Penang Port ranked the third busiest container port in Malaysia after port Klang and the port of Tanjung Pelepas [6]. In 2017, the port of Penang was Malaysia's third busiest port in terms of cargo, handling 1.52 million TEUs, and the country's busiest port-of-call for cruise ships, overtaking port Klang in 2017. The Swettenham Pier

A. F. Ahmad Fuad (✉) · N. I. I. Zamri · A. S. Kamis
Nautical Science and Maritime Transportation Program, Faculty of Maritime Studies, Universiti Malaysia Terengganu, Kuala Nerus, Terengganu, Malaysia
e-mail: faizalfuad75@gmail.com

N. I. I. Zamri
e-mail: imaninsyirah97@gmail.com

A. S. Kamis
e-mail: amir.syawal@umt.edu.my

M. S. I. Saadon
Maritime Management Program, Faculty of Maritime Studies, Universiti Malaysia Terengganu, Kuala Nerus, Terengganu, Malaysia
e-mail: saiful.izwaan@umt.edu.my

A. Fitriadhy et al. (eds.), *Advancing Interdisciplinary Knowledge and Innovation*,
SpringerBriefs in Applied Sciences and Technology,
https://doi.org/10.1007/978-3-032-01309-5_9

Cruise Terminal plays a vital role as a major entry point for tourists into Penang, aside from the Penang international airport. The level of risk of vessel collision along Penang Port's navigation channel is after port Klang and the port of Tanjung Pelepas [7].

Problem Statement and Objectives

The Penang Port lacks a radar-equipped VTS station, relying on AIS for limited vessel tracking. This study aims to establish a VTS station by examining relevant standards and guidelines. Key research objectives include.

i. To identify technical standards and guidelines to establish a VTS station for the port; ii. To apply the selective technical standards and guidelines; iii. To propose the best site for the VTS station.

This study seeks to allow the guideline on technical standards applied using radar range and VHF range calculation to determine a suitable location at Penang Port to establish a VTS station at which terminals should be encouraged and suitable to apply the VTS station.

2 Methodology

The overall research activities are shown in the flowchart of Fig. 1. A detailed explanation is given in the following paragraphs. Step 1 involves obtaining the VTS technical standard, which is outlined in the IALA guidelines for establishing a VTS station for ports. The standards are searched for using two methods: firstly, through online databases like Science Direct and Google Scholar using the keywords "Vessel Traffic Services," and, secondly, through the IALA website. Articles are selected in three stages: by title, abstract, and full article. However, only two out of thirteen technical requirements are relevant for researching the establishment of a vessel traffic system: Radar and radio communication (VHF).

Step 2 involves identifying the requirements for VTS at Penang Port through interviews with Penang Port Sdn. Bhd. management. Interviews were conducted with two experienced experts from the marine department familiar with maritime operations at Penang Port. Interview questions were based on the IALA guideline "Preparation of Operational and Technical Performance Requirements for VTS Systerms" [4]. The needed requirements are as follows:

i. Delineating the VTS area; ii. Types of services to be provided (information services, traffic organization services, navigational assistance services); iii. Types and sizes of vessels which are required or expected to participate in the VTS; iv. Navigational hazards and traffic patterns; v. Human factors including health and safety issues; vi. Tasks to be performed by system users.

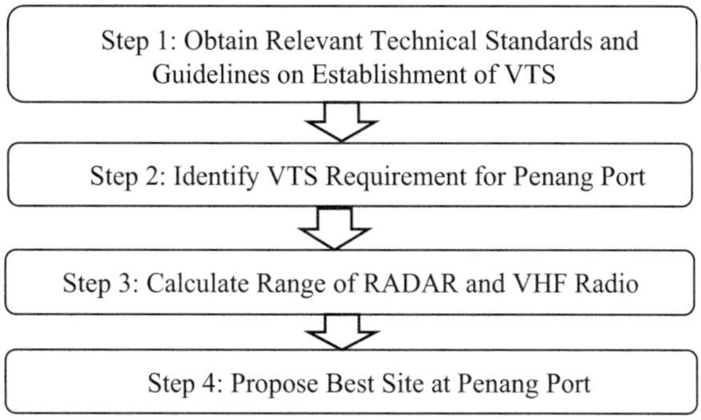

Fig. 1 Flowchart of research activities

Step 3 is to calculate the radar and VHF radius range at selected sites in Penang Port. The sites were NBCT and the Swettenham pier cruise terminal. The coordinates are shown in Table 1. The location of Swettenham pier cruise terminal and Penang Port's building is shown in Fig. 2. The calculation of maximum radius for radar was using Eq. (1) [8]. The calculation of the maximum radius of the VHF radio was using Eq. (2) [9]:

$$R_{\text{Max}} = \left[\frac{P_t G^2 \lambda^2 \sigma}{(4\pi)^2 S_{\min}} \right]^{1/4} \tag{1}$$

where

R_{Max} = maximum radar range; P_t = transmitted power by radar; G = maximum power gain of antenna; λ = operating wavelength; σ = radar cross section; $S_{\min}$ = power of minimum detectable signal:

$$d = 4.1.2 \times \sqrt{h} \tag{2}$$

where

d = distance in kilometre and h = height of antenna in meter.

Step 4: To propose the best site at Penang Port. Two sites were proposed for the VTS station, namely Penang Port's building and the Swettenham pier. Results on the calculation of the radar and VHF radio range would be used to determine the site. Suitability of the site to place the radar scanner and VHF radio antenna was determined to obtain optimum performance.

Table 1 Name and coordinates of Penang Port terminal

Name of the terminal/station	Coordinates
Northern Butterworth Container Terminal	5° 24′ 35″ N, 100° 22′ 02″ E
Swettenham Pier Cruise Terminal	5° 25′ 08″ N, 100° 20′ 43″ E

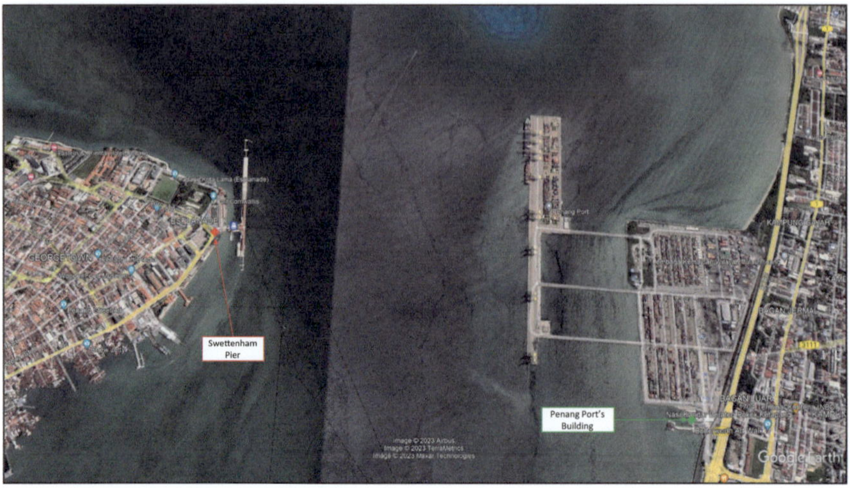

Fig. 2 Location of Swettenham pier cruise terminal (red dot) and Penang Port's building (green dot)

3 Results and Discussion

Relevant Technical Standards, Guidelines, and Recommendation for VTS Establishment

There are three documents that are relevant for the establishment of VTS in a particular port or sea area. IALA Guidelines G1111, G1150, and G1089 outline frameworks for VTS systems. G1111 focuses on system development, G1150 aids in implementing VTS practices, and G1089 offers standardized service delivery guidance. G1150 is particularly relevant for setting up VTS stations.

Results of VTS Requirement for Penang Port

The semi-structured interview questionnaire results regarding radar coverage of the VTS Penang area were distributed to both experts at Penang Port. Table 2 presents the questions and answers from these experts, developed based on IALA Guideline G1150. The questions aim to understand operational requirements for radar coverage, including preferred coverage, target detection, environmental conditions, influencing factors, target detection requirements, separation, positional accuracy, required update rate, and opinions on radar dynamic capabilities and constraints.

Table 2 Response from the experts

Question	Expert 1	Expert 2
What is the preferred radar coverage of the VTS Penang area?	24 nm	24 nm
What is the target/object to be detected at Penang Port?	• Container carriers, tankers, etc. • Large coasters, bulk carriers, cargo ships, etc. • Coasters, etc. • Small metal ships, fishing vessels, patrol vessels, etc. • Sailing boats, speedboats, etc. • Fiberglass, wood, or rubber boats with outboard motors, at least 4 m long	• Container carriers, tankers, etc. • Large coasters, bulk carriers, cargo ships, and the like • Coasters and the like • Small metal ships, fishing vessels, patrol vessels, and the like
What determines environmental capabilities and constraints? What are your preferred maximum working conditions regarding hourly precipitation rate?	Moderate 11–30 (mm/hr)	Moderate 11–30 (mm/hr)
What are the other factors, obstructions, and interference present at Penang Port?	• Influenced by topography • Influenced by buildings and other man-made structures like quay crane and bridge	• Influenced by topography • Influenced by buildings and other man-made structures
What are the minimum target detection requirements under typical weather conditions in the VTS area?	Small 4m LOA fiberglass boat during normal rain	Class C fishing vessel during clear weather
What are the target separation and positional accuracy?	5–20 NM range	5–20 NM range
What is the required radar update rate?	Between 2 and 4 s	Between 2 and 4 s
What are the radar dynamic capabilities and constraints?	The right location and height	Penang Port's vast pilotage zone necessitates multiple radar installations due to obstructed views. Acquiring land, like Rimau Island, is necessary. High-speed internet is vital for real-time data at the port control center. However, costs outweigh vessel traffic's financial impact

Table 3 Input radar range calculation

Symbols	Parameter	Data
P_t	The maximum amplitude of transmitted power, watts	25 W
G	Gain of transmitting antenna	30 dB
λ	Wavelength	0.032 m
σ	The cross-sectional area of the object, m^2	20.1 m^2
S_{min}	The minimum amplitude of the signal that can be detected by the antenna at the receiving end, watts	1.0^{-16} W

Calculation of Marine Radar Range

The calculation of radar range for Penang Port is based on Kuantan port VTS radar, namely SIMRAD Halo20+ x-band marine radar. The inputs to Eq. (1) are shown in Table 3. Results of the $R_{\max}$ calculation for the 25 W maximum transmitted power and Class A fishing vessel radar cross section is 7 NM or 13.1 km. This result is not up to the requirement set by the experts from Penang Port. Therefore, adjustment in the radar's parameters should be done. The radar's maximum transmitted power was changed to 3.5 kW to acquire the maximum distance of 24 NM or 45 km for detection of a Class A fishing vessel. The radar distance of 24 NM is shown in Fig. 3, and this distance had covered beyond Penang Port's limit and north pilot station. A larger commercial vessel would be detected more than 24 NM.

Calculation of VHF Radio Range

Using Eq. (2), the VHF radio calculation utilized the height of the Penang Port Sdn. Bhd. building, which stands 34.4 m tall. This results in a maximum VHF radio range of 13 NM or 24.16 km radius from the building (Fig. 3), covering the south channel but falling short at the north channel. To ensure communication with incoming ships near buoy number 1 in the north channel, the VHF radio range should be increased to at least 20 NM or 37 km from the building. Achieving this requires a VHF radio antenna height of at least 81 m, as per Eq. (2). However, using higher transmission power can compensate for a shorter antenna height while maintaining the desired range.

The Best Site for VTS Station in Penang Port

Two sites were evaluated at Penang Port: Swettenham pier and Penang Port's building. However, the presence of large cruise ships like the Spectrum of the Seas at Swettenham pier could obstruct VTS radar and VHF radio transmissions. As the

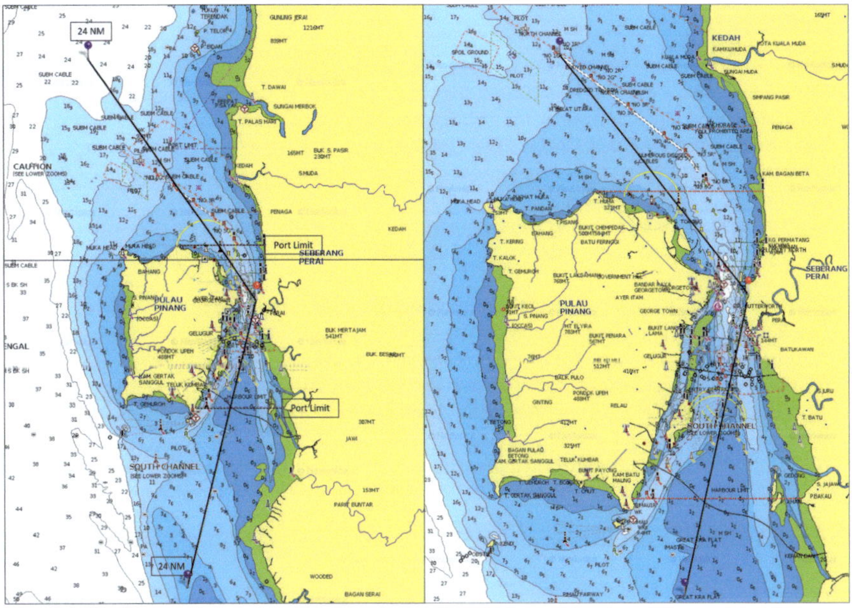

Fig. 3 Radar range of 24 NM from Penang Port's building (left) and VHF radio range of 13 NM from top of Penang Port's building (*Source* Navionics (2023))

pier can host two such ships concurrently, Penang Port's building is deemed a more suitable location for the VTS station.

4 Conclusion

This study had three objectives: providing technical guidelines for VTS establishment, gathering input on VTS requirements from Penang Port experts, and calculating marine radar and VHF radio ranges. To achieve the required ranges, radar transmitting power was increased, and VHF antenna height was adjusted. Ultimately, Penang Port's building was chosen as the VTS station location due to fewer transmission blockages compared to Swettenham pier.

Acknowledgements The research team would like to thank the Holistic Lab of University Technology Malaysia, which had partly the finance of this research.

References

1. IMO, *Resolution A.1158(32) Guidelines for Vessel Traffic Services* (London, 2021)
2. IALA, *Navguide 2018 Marine Aids to Navigation Manual* (2018)
3. IMO, *SOLAS Consolidated Edition 2009* (IMO, London, 2009)
4. IALA, *IALA Guideline G1111 Preparation of Operational and Technical Performance Requirements for VTS Systems* (2022)
5. G. Praetorius, M. Lützhöft, Decision support for vessel traffic service (VTS): user needs for dynamic risk management in the VTS. Work **41**, 4866–4872 (2012). https://doi.org/10.3233/WOR-2012-0779-4866
6. E. van der Heide, *Port Development in Malaysia an Introduction To the Country's Evolving Port Landscape* (Netherland, Amsterdam, 2019)
7. A.F.A. Fuad, M.K. Hatun, A.S. Kamis, A.M. Dom, M.S., Ahmad, Study on traffic and probability of collision at navigation channel of the top three ports in Malaysia (2023)
8. Tutorials Point, *Radar Systems—Range Equation Derivation of Radar Range Equation*, https://www.tutorialspoint.com/
9. M. Brown, *VHF Radio Range Calculator*, https://3roam.com/vhf-radio-range-calculator/

Development of Smart Mushroom Farming System Using IoT

Muhammad Ilyas Zulkhairi, Nurul Atiqah Hasman, Izanoordina Ahmad, Nor Hidayah Abdul Kahar, Siti Marwangi Mohamad Maharum, and Zuhanis Mansor

1 Introduction

Oyster mushrooms, also known as Pleurotus Ostreatus, are a type of edible mushroom that is commonly cultivated for consumption [1]. They are known for their delicate flavor and texture and are often used in a variety of dishes, including soups, stir-fries, and pasta. Growing oyster mushrooms requires a few specific parameters, including the right temperature, humidity, and substrate. The ideal temperature range for growing oyster mushrooms is in the range 60–70 degrees Fahrenheit, with the humidity level of around 80–90%. The substrate, or growing medium, for oyster

M. I. Zulkhairi · N. A. Hasman · I. Ahmad (✉)
Intelligent Embedded Research Lab, Electronics Technology Section, British Malaysian Institute, Universiti Kuala Lumpur, Gombak, Selangor, Malaysia
e-mail: izanoordina@unikl.edu.my

M. I. Zulkhairi
e-mail: ilyaszulkhairi@gmail.com

N. A. Hasman
e-mail: natiqah.hasman@s.unikl.edu.my

N. H. A. Kahar
Electrical Technology Section, British Malaysian Institute, Universiti Kuala Lumpur, Gombak, Selangor, Malaysia
e-mail: norhidayahkahar@unikl.edu.my

S. M. M. Maharum · Z. Mansor
Advanced Telecommunication Technology, Electronics Technology Section, British Malaysian Institute, Universiti Kuala Lumpur, Gombak, Selangor, Malaysia
e-mail: sitimarwangi@unikl.edu.my

Z. Mansor
e-mail: zuhanis@unikl.edu.my

A. Fitriadhy et al. (eds.), *Advancing Interdisciplinary Knowledge and Innovation*, SpringerBriefs in Applied Sciences and Technology,
https://doi.org/10.1007/978-3-032-01309-5_10

Table 1 Requirements of Oyster mushroom

Oyster mushroom parameters		
No	Measure	Travel direction
1	Temperature	28–30 °C
2	Humidity	80–90%
3	Carbon dioxide	1200–1500 ppm

mushrooms can be made from a variety of materials, including straw, sawdust, and even coffee grounds [2].

One of the main difficulties of mushroom farming is controlling the environment. Mushrooms are highly sensitive to temperature, humidity, and light levels [3]. The mold grows in places with high moisture and can damage the crop, which explains why temperature and humidity control are important. Table 1 shows the requirements for growing oyster mushrooms. In traditional mushroom farming, growers must manually monitor and adjust these factors, which can be time-consuming and labor-intensive, which can increase costs and reduce efficiency. The research in [4] discussed the application of IoT in agriculture, specifically in the context of "smart agriculture." The aim of IoT-based smart agriculture is said to enhance the agricultural sector by monitoring the field in real-time using sensors and interconnectivity, which can save farmers time and reduce the wastage of resources.

The research by Kanchana [5, 6] discussed the use of IoT in smart farming. Technology advancements have enabled farmers to improve crop production through IoT integration. Smart farming, using wireless monitoring sensors and a cloud-based platform, can detect soil quality, meteorological conditions, crop growth, and crop damage remotely, thus boosting agricultural production at a lower cost and more efficiently. The research in [7] discussed the use of technology in irrigation management in agriculture. The importance of precise irrigation in boosting crop yields and saving resources is highlighted in the article. Recent advancements in technology have led to the introduction of advanced automated irrigation methods that use sensors, actuators, and wireless networks to form a closed-loop irrigation system, which optimizes water use and reduces water wastage, labor cost, and time.

Thus, this project will be using IoT in mushroom farming by automating the monitoring and control of growing conditions and reducing the need for manual labor. By using sensors and other Internet-connected devices, mushroom farmers can continuously monitor and optimize the conditions for growing mushrooms. This leads to increased efficiency and crop yield with improved quality.

2 Methodology

Figure 1 shows the block diagram of the mushroom farming system. The block diagram of the project is divided into three main parts: inputs, data processing, and output. The input of the project includes an AM2303 temperature and humidity sensor, as well as an MQ135 sensor, which will collect data on temperature, humidity, and CO_2 levels in the mushroom growing environment. This data is then sent to the ESP32 for processing via a wired connection and streamed to the Favoriot platform for monitoring. If certain threshold values are exceeded, the system will trigger alerts for the user through Telegram. The output of the project includes a 12 V PC fan, humidifier, LCD display, and LED strip. These components are used to control the temperature and humidity in the growing environment and are instructed to function based on the data collected by the sensors. The LCD display will display important parameters such as temperature, humidity, and CO_2, and the LED strip can be turned on by the user to trigger mushroom pin growth.

Figure 2 shows the flowchart for the process of checking mushroom parameters. The flowchart illustrates the steps of the process using various shapes. The process begins by initializing the system and calibrating the sensors. The AM2303 temperature humidity sensor and MQ135 gas sensor then gather data on temperature, humidity, and CO_2 levels in the mushroom environment and send it to the ESP32 for processing. If the temperature exceeds 30 degrees Celsius, an alert is sent to the user via Telegram, and the fan is turned on until the temperature drops below 30°. The system then checks the humidity level and, if it is below 70%, sends another alert to the user via Telegram and activates the humidifier until the humidity level increases above 70%. Lastly, the system checks for CO_2 levels and, if it exceeds 1500 ppm, sends an alert to the user via Telegram and turns on the fan to exhaust the excess CO_2.

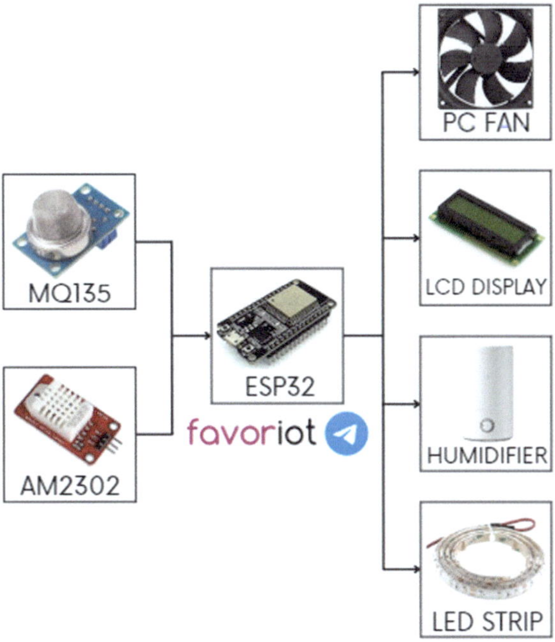

Fig. 1 Block diagram of the system

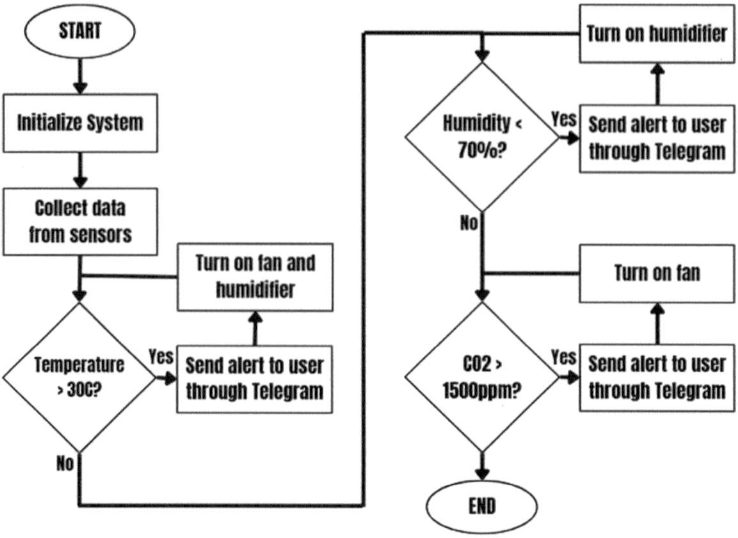

Fig. 2 Process flowchart for checking mushroom parameters

3 Results and Discussion

Figure 3 shows the circuit connection of the project. The sensors will send the data to ESP32. The data is displayed on the LCD and Favoriot platform. Figure 4 shows the data displayed on the LCD. The relays are used to control the outputs fan, humidifier, and LED strip as shown in Fig. 5. External power supply 12 V and SV are used to power up the outputs through the use of relays. Besides that, Fig. 6 illustrates the data streamed to Favoriot. When the sensor has read the temperature, humidity, and CO_2 level in the container, ESP32 will stream the information over the Wi-Fi to the Favoriot platform. The data can be represented in graphs for monitoring. Figure 7 shows the graphs for the factors versus time. The ability of ESP32 is to communicate with the sensors and also verify the functionality of the sensors that are successful based on the data displayed. The sensor testing displays real-time information from the temperature, humidity, and gas sensors. Streaming of data to Favoriot illustrates the ability to connect to the IoT platform. Using Favoriot, data can be monitored from remote locations and users can be alerted of unfavorable growing conditions.

The performance of the system is tested by observing the growth of mushrooms in the container. Figure 8 shows the placement of mushroom blocks in the prototype. The LED strip is turned on to trigger the growth of the mushroom pin. Once the pin has grown, the user can switch off the LED by pressing the push button. The user will be alerted through Telegram when a specific threshold is exceeded. Figure 9 shows the alert sent by Favoriot to the user through Telegram.

Fig. 3 Circuit connection

Fig. 4 Data display on LCD

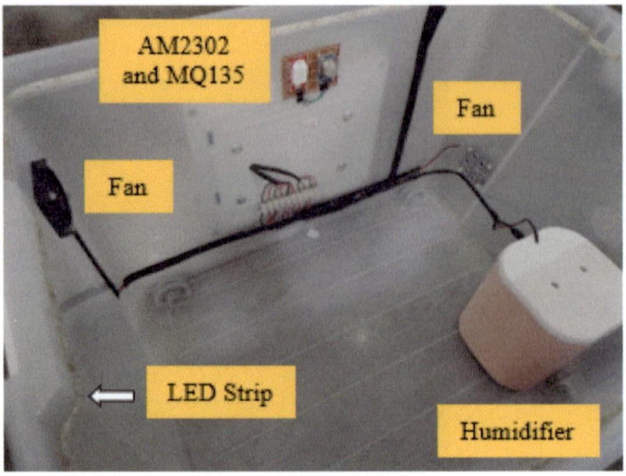

Fig. 5 Internal view of project

Select	Device Developer ID	Data	Uploaded	
☐	MushroomSensors@izanoordina	'co2':'401','humidity:'60','temperature':'30'	1/4/2023, 11:19:05 AM	
☐	MushroomSensors@izanoordina	'co2':'401','humidity:'64','temperature':'29'	1/4/2023, 11:11:05 AM	
☐	MushroomSensors@izanoordina	'co2':'401','humidity:'76','temperature':'29'	1/4/2023, 11:11:06 AM	
☐	MushroomSensors@izanoordina	'co2':'401','humidity:'76','temperature':'28'	1/4/2023, 11:11:35 AM	
☐	MushroomSensors@izanoordina	'co2':'401','humidity:'78','temperature':'30'	1/4/2023, 11:11:05 AM	
☐	MushroomSensors@izanoordina	'co2':'401','humidity:'72','temperature':'30'	1/4/2022, 11:11:01 AM	
☐	MushroomSensors@izanoordina	'co2':'401','humidity:'72','temperature':'29'	1/4/2023, Activate W30:59AAM s	

Fig. 6 Data stream to favoriot

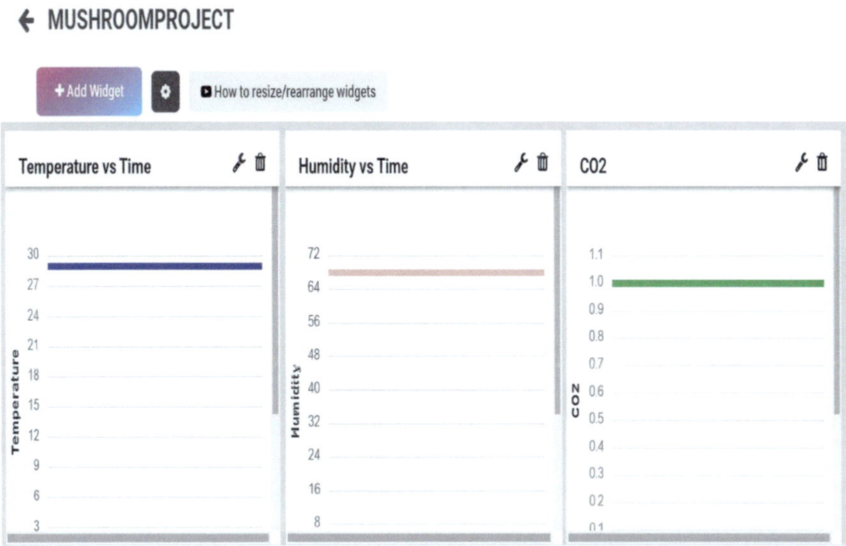

Fig. 7 Representation of data in graphs

Fig. 8 Mushroom blocks in prototype

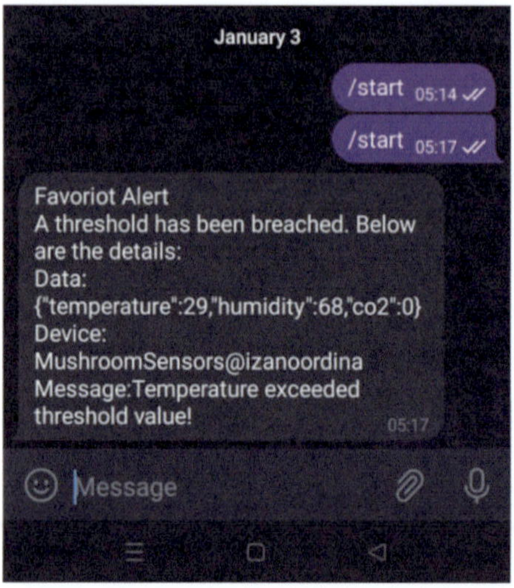

Fig. 9 Alert from telegram

4 Conclusion

In conclusion, the increasing global population and advancement of technology led to the adoption of the Internet of Things in the agricultural industry. The design of the system for mushroom farming using IoT was successful. The project can collect important parameters of the mushroom and stream it online to the Favoriot platform. In the future, the system will be carried out in mushroom farms to gauge the performance of the system. For future improvement of the project, the virtual button should be added for easier control of LED. Additionally, a camera module can be installed into the container to capture the growth of mushrooms in real time. The data collected can be used in machine learning algorithms to predict and optimize mushroom growth, by analyzing sensor data and adjusting the environment as needed. Overall, the sensors and actuators in this project work as intended according to the flowchart.

Acknowledgements This research was conducted by the students and staff of Universiti Kuala Lumpur British Malaysian Institute (UniKL BMI), and UniKL financially supports its publication. Special thanks to the Intelligence Embedded Research Lab (IERL) for providing the lab throughout the project.

References

1. C. Muswati et al., The effects of different substrate combinations on growth and yield of Oyster mushroom (Pleurotus ostreatus). Int. J. Agron. (2021)

2. A. Najmurrokhman et al., Development of temperature and humidity control system in internet-of-things based Oyster mushroom cultivation. In: *2020 3rd International Seminar on Research of Information Technology and Intelligent Systems, ISRITI* (2020)
3. N.P. Shah, Greenhouse automation and monitoring system design and implementation. Int. J. Adv. Res. Comput. Sci. (2017)
4. S. Kanchana, IoT in agriculture: smart farming. Int. J. Sci. Res. Comput. Sci. Eng. Inf. Technol. (2018)
5. A. Subedi et al., IoT Based monitoring system for white button mushroom farming. In: *6th International Electronic Conference on Sensors and Applications session Sensor Networks* (2019)
6. S.V. Kumar et al., Review on IoT based precision irrigation system in agriculture. Curr. J. Appl. Sci. Technol. (2020)
7. R. Krishna Tummala et al., Sustainable farming for agriculture improvement using IoT. Int. J. Modern Agric. (2021)

Evaluation of Implementation of Occupational Health and Safety Procedures for Airport Operations at PT Angkasa Pura I Sultan Hasanuddin International Airport Post COVID-19 Pandemic

Akbar Fadillah, Muhammad Rafly Ananta, Datunabolon, and Retno Mawarti S. Sos

1 Introduction

Air travel's contribution to global mobility is enormous, and its operation must comply with strict safety and security standards. The current pandemic crisis has had a significant impact on aircraft operations, requiring modifications to standard operating procedures to take into account future sanitation needs. PT. Angkasa Pura 1 (Persero) is an airport processing services company. Sultan Hasanuddin International Airport (UPG) is one of the airports under the management of PT. Angkasa Pura I (Persero). Sultan Hasanuddin Airport is one of the entry points for local and international tourists to South Sulawesi, and this airport is considered as one of the airports that implements occupational safety and health programs. He revealed that PT. Angkasa Pura 1 (Persero) Sultan Hasanuddin Makassar International Airport received the 2019 Gold Flag Award with a level of occupational safety and health implementation reaching 92.16% [1].

A. Fadillah (✉) · M. Rafly Ananta · Datunabolon · R. M. S. Sos
Trisakti Institute of Transportation and Logistics, East Jakarta City, Indonesia
e-mail: fadillahjunior23@gmail.com

M. Rafly Ananta
e-mail: rafly.ananta07@gmail.com

Datunabolon
e-mail: danutabolon@gmail.com

R. M. S. Sos
e-mail: retno_03@gmail.com

A. Fitriadhy et al. (eds.), *Advancing Interdisciplinary Knowledge and Innovation*, SpringerBriefs in Applied Sciences and Technology,
https://doi.org/10.1007/978-3-032-01309-5_11

The occupational safety and health management system which is applied at PT. Angkasa Pura I (Persero) Makassar Sultan Hasanuddin International Airport was awarded the Golden Flag, but that does not mean it is running effectively. The transmission of COVID-19 cases increased significantly in January 2021, which may be due to the lack of self-control and awareness of each individual about health regulations. The spread of this virus does not rule out the possibility of attacking workplaces in large numbers, including airport operations. So it is very important to improve the occupational health and safety program to stop the spread of COVID-19 in the workplace. This study aims to examine and evaluate the implementation of the occupational safety and health program at Sultan Hasanuddin International Airport in an effort to prevent the spread of the COVID-19 virus from increasing again.

Work safety includes protecting people while doing their jobs. This applies to both the manufacturing sector using working equipment, materials, and processing processes and the service sector using advanced technological equipment such as elevators, escalators, building cleaning equipment, and means of transportation [2]. According to another personification, workplace safety is a condition in which workers feel comfortable and safe, affecting the level of output and quality of work. Efforts to prevent occupational accidents and occupational diseases include the implementation of technical measures to secure workspaces, machinery, equipment, and the working environment as well as administrative controls [3].

Occupational health is the absence of physical and mental disorders caused by this work. In the workplace, occupational health must be considered because people who work effectively are often in good health. On the other hand, if a worker's health is disturbed due to illness, accident, or disability, it will negatively affect performance at work, thereby reducing work efficiency [2]. When considering occupational health and safety, this relates to the organization's efforts to protect employees from the risk of accidents, illnesses, and losses arising from the performance of work.

According to [4], occupational health and safety refers to the management of people and objects, including methods related to the work environment so that workers do not become injured. System of programs called occupational health and safety has been established for workers and employers with the aim of reducing the incidence of illness and accidents in the workplace. In [5], it is explained that occupational health and safety is a program system developed for workers and employers to prevent work accidents and work-related diseases caused by work relations in the work environment, as well as identifying factors that can cause work accidents and taking proactive action to respond in that case.

In [6], it is explained that airports are the gateway to a region or country, and are the main means of the air transportation sector. In order for an airport to carry out its functions, it must be organized in an integrated manner it must be organized in an integrated manner so that it can provide effective services to customers. An airport is a specific location on land or water (including buildings, facilities, and equipment) designed in whole or in part for the arrival, departure, and movement of aircraft, in accordance with ICAO Annex 14.

2 Methodology

Quantitative research methods are used in this study. Primary and secondary data were collected for this study through observation and interviews. Researchers distributed questionnaires to employees of PT. Angkasa Pura I to collect data. Questionnaires distributed to respondents became a source of secondary data for the study, and statistics on the increase in cases of the COVID-19 virus at Sultan Hasanuddin Airport became a source of secondary data for the study. The SPSS version 26 program is used by researchers and can assist in the fast and correct processing of research data results. The validity test and the reliability test are used by researchers to test the quality of the data, all of which must be passed by the researcher so that the research is considered valid. Verification of the accuracy is done to determine how much variation is obtained and how this reflects inequality among study respondents. Reliable questionnaires have answers that remain constant over time.

1. In this research, the results of the T-test indicate that work safety has an effect on airport operations from Table x, the value of Tcount is 1.684 where Tcount < Ttable (1.684 < 1.984) with a sig level value of 0.095 > 0.05; this is very influential on airport operations, and we can conclude that the work safety is accepted.
2. In the occupational health variable, the Tcount value is 6.034 and Ttable (6.034 > 1.984) with a sig level of 0.00 < 0.05, then Table y shows that the Occupational Health factor has no effect on airport operations.

This F-test Anova in Table 2 shows that the Fcount value is 39.394 while the Ftable value is 3.09. With a significant value of 0.000 < 0.005, the Fcount value is 39.394 > 3.09 (dfl = 2, df2 = 97). So, it can be said that the variables of occupational safety and health have a significant influence on the operation simultaneously because there is no rejection and acceptance [7].

Table 1 Test T

Coefficients[a]

Model		Unstandardized coefficients		Standardized coefficients	T	Sig.
		B	Std. error	Beta		
1	(Constant)	6.834	2.582		2.647	0.009
	Keselamatan Kerja	0.106	0.603	0.157	1.684	0.095
	Kesehatan Kerja	0.655	0.109	0.564	6.034	0.000

[a]*Dependent Variable* Operational Bandar Udara

Table 2 Test F

ANOVA[a]						
Model		Sum of squares	df	Mean square	F	Sig.
1	Regression	204.487	2	102.244	**39.394**	0.000[b]
	Residual	251.753	97	2.595		
	Total	456.240	99			

[a]*Dependent Variable* Operational Bandar Udara
[b]*Predictors* (Constant), Kesehatan Kerja, Keselamatan Kerja

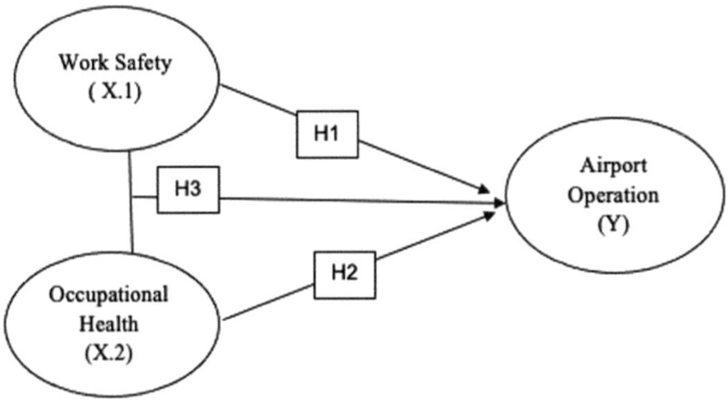

Fig. 1 Conceptual framework

3 Conclusion

1. Work safety has a positive and significant effect on the airport operations of PT. Angkasa Pura I Sultan Hasanuddin International Airport, Makassar. The results of Hypothesis 1 states that work safety has a positive effect on airport operations, which means that work safety is within the scope of PT. Angkasa Pura I Sultan Hasanuddin Airport Makassar that has been running according to the existing Standard Operating Procedures and because H1 was accepted.
2. Occupational health has no significant effect on the operations of PT. Angkasa Pura I Airport Sultan Hasanuddin International Airport Makassar, so Hypothesis 2 is rejected which means occupational health is not in accordance with tbe existing standard operational procedures; this should require a thorough health supervision of these employees and cleanliness in the scope of airport operations of the airport operation.
3. Occupational safety and health have a positive and significant effect on airport operations; the results of Hypothesis 3 show that occupational safety and occupational health together (simultaneously) have a significant effect on airport operations. With the implementation of good SMK3, airport operations will run well too; this must be supported by a thorough supervision and audit to ensure that the

SMK3 that is being implemented is running according to the existing standard operational procedures (Fig. 1 and Tables 1 and 2).

Acknowledgements Therefore, we express our deepest gratitude to all those who have helped and supported the author in the preparation and completion of this journal, namely to:

1. Mr. Dr. Tjuk Sukadirman, as Chancellor of the Institute of Transportation and Logistics.
2. Mrs. Marthaleina R. S., S. Kom, MM, as Dean of the Faculty of Management and Business Dean of the Faculty of Management and Business.
3. Mr. Datunabolon, ST, MM, and Mrs. Retno Mawarti S. Sos, MM, as supervisors who have spent a lot of time, energy, and thoughts in guiding and directing the author in completing this journal.
4. Thanks to the employees of PT. Angkasa Pura I Sultan Hasanuddin International Airport Makassar who has been wıllıng to help the author, take the time between his busy schedules, and assist the author in filling out and distributing questionnaires. Without your help, this journal would not have been completed on time.

References

1. K. Lucky Afri Senjani et al., Faktor Yang Mempengaruhi Perilaku Karyawan Terhadap Penerapan Smk3 Di Pt. Angkasa Pura I (Persero). Wind. Public Heal. J. **01**(06), 702–714 (2021)
2. R. Hasibuan, Pengaruh Kesehatan Dan Keselamatan Kerja, Pelatihan Dan Kerja Tim Terhadap Kinerja Tenaga Medis Di Rumah Sakit Budi Kemuliaan Batam. J. Dimens. **6**(2), 323–340 (2017). https://doi.org/10.33373/dms.v6i2.1054
3. R.D. Kartikasari, B. Swasto, Pengaruh keselamatan dan kesehatan kerja karyawan terhadap kinerja karyawan. Adm. Bisnis **44**(1), 1–7 (2017)
4. V.K. Nan Wangi, Dampak Kesehatan Dan Keselamatan Kerja, Beban Kerja, Dan Lingkungan Kerja Fisik Terhadap Kinerja. J. Manaj. Bisnis **7**(1), 40–50 (2020). https://doi.org/10.33096/jmb.v7i1.407
5. R. Hadiyanti, M. Setiawardani, Pengaruh Pelaksanaan Program Keselamatan dan Kesehatan Kerja Terhadap Produktivitas Kerja Karyawan. J. Ris. Bisnis dan Investasi **3**(3), 12 (2018). https://doi.org/10.35697/jrbi.v3i3.941
6. M.A. Wilman, Pengaruh Kualitas Pelayanan Terhadap Kepuasan Pelanggan Terminal 3 Bandara Soekarno-Hatta (2017), p. 156
7. S. June, M. Siagian, Pengaruh Keselamatan Dan Kesehatan Kerja Dan Lingkungan Kerja Terhadap Kinerja Karyawan Pt Lautan Lestari Shipyard. J. Apresiasi Ekon. **8**(3), 407–420 (2020). https://doi.org/10.31846/jae.v8i3.325

Critical Success Factors for Improvement Initiative Deployment

Faris Hakim Mazlan and Mohamad Ikbar Abdul Wahab

1 Introduction

Improvement initiative refers to any improvement activities, methods, or programs implemented by organizations in order to achieve their improvement objectives [1], which consist of multiple levels [2] such as methodology, tools, and/or techniques [3]. Based on the definition and classification level of IMI stated by [4], as shown in Fig. 1, the scope of this study is at the strategic level, focusing on the initiatives or methodologies themselves rather than their tools and techniques. Based on the references of previous studies, the list of strategic initiatives, which are also commonly reported in the annual report of the Malaysian Productivity Corporation (MPC) [5] are Lean, Six Sigma, TRIZ, ISO 9000, BPR, Total Quality Management (TQM), ICC/QCC, Lean Six Sigma, and Business Excellence.

For the past three decades, a number of different approaches have gained huge interest and applications in the industry world such as Total Quality Management (TQM), Six Sigma, Lean, and BPR [6]. One thing remains, however, that assuring the success of the relevant IMI deployment relies on the foundation that has been viewed as the basis for determining the information requirements of corporate managers [7], which is called the critical success factor (CSF)—a term that was popularized by John F. Rockart in his 1979 paper which represents the essential ingredients in which without it, an initiative has a low probability of success [8].

Companies continue to strategize in order to withstand the intense competition and ultimately strike with IMI. Moreover, as a result of their haphazard approach

F. H. Mazlan (✉) · M. I. A. Wahab
Quality Engineering, Malaysian Institute of Industrial Technology, Universiti Kuala Lumpur, Johor Bahru, Malaysia
e-mail: farishakim366@gmail.com

M. I. A. Wahab
e-mail: mikbar@unikl.edu.my

A. Fitriadhy et al. (eds.), *Advancing Interdisciplinary Knowledge and Innovation*, SpringerBriefs in Applied Sciences and Technology,
https://doi.org/10.1007/978-3-032-01309-5_12

to survival, they failed to prioritize the IMI success criteria during implementation, resulting in failure. Several researchers found, through a review of the existing literature, that a large proportion of organizations that implement improvement methodology fail to survive [9] and that around 60% of Six Sigma initiatives fail to produce the desired outcome. Others, such as [10], indicate that fewer than 10% of UK organizations are successful in their Lean implementation efforts, whereas the failure rate of TQM implementations is comparable to that of other strategies [11], with success rates reported to range from 10 to 30% in Europe [12], in which one of the causes of failure is due to the lack of CSF awareness [13]. Organizations that are ignorant of or lax about the CSF of IMI will fail because they did not adhere to the criteria that lead to IMI's success.

2 Research Methodology

This research was divided into three main phases, beginning with a literature search using SLR, followed by the building of a conceptual model, then the validation of the model using systematic analysis techniques, and concluding with a summary of the study's findings, as per Fig. 2. For Phase I, a systematic literature review was conducted in accordance with the Preferred Reporting Items for Systematic Reviews and Meta-Analyses (PRISMA) standard, identifying 2194 publications that were shortlisted down to 72 papers as shown in Fig. 3. The information extracted through SLR was used to develop the conceptual model of CSFs for strategic IMI, which is explained in Phase II. After reviewing and compiling all critical success factors associated with nine strategic initiatives, the development of the CSF model for IMI continues with the clustering and grouping of CSFs using an affinity diagram, one of the seven new quality tools. This quality tool facilitates the separation of all mentioned CSFs as constructs along with their items (attribute) into an appropriate group, allowing for the establishment of a new group heading that justifies a systematic manner of CSF and its attribute. Once the conceptual models has been established; it is validated by IMI experts through the distribution of a survey questionnaire to various IMI-using enterprises.

Once the survey data is retrieved, it needs to get processed to make it useful. For the ease of workflow for this analysis, the authors divided the process into two stages. Stage I is by utilizing Statistical Package for the Social Sciences (SPSS) and Stage II via Smart PLS to conduct structural equation modeling (PLS-SEM).

3 Result and Discussion

Based on 72 reviewed papers, a total of 239 attributes for the combined nine initiatives were mentioned in previous research. These lists of information are then used in developing a conceptual model of CSF for strategic IMI implementation.

After compiling all 239 attributes of the success factors, the extracted data is then used to generalize the factors under one conceptual model. Several of the success factors identified in prior research have been cited by multiple writers with slight variations in wording, rendering the attributes redundant, while others stand alone. Due to that reason, the authors have difficulty in rearranging the factors to the appropriate group of constructs and items. Thus, the affinity diagram (one of seven new quality tools) was utilized to cluster the construct and attributes systematically. By using the tool to categorize and map the related attributes to its individual construct correctly, the author managed to categorize the CSFs associated with those nine initiatives and generalized them into one conceptual model which consists of nine CSFs and 40 attributes. Figure 4 shows the comprehensive model of critical success factors for the IMI.

Next, the validation of the developed model is done through the administering of the mass survey in which the collected responses were uploaded into SPSS, to conduct data screening, normality test, non-response bias test, common method bias test, and descriptive statistics. Subsequent to the SPSS analysis, structural equation modeling (SEM) analysis was then conducted [14] in which the model is based on a reflective-formative type hierarchical latent variable. In this analysis, there are two parts involved. For Part A, three main steps were involved in measuring the reflective model where these steps were done to test the internal reliability (IR), convergent validity (CV), and the discriminant validity of the developed model. Based on these three data analysis results, the measurement model for the second-order reflective-formative model (internal consistency, convergent validity, and discriminant validity) fulfilled all required criteria. Thus, the measurement model of critical success factors is all reliable and valid, which then proceeds to structural model evaluation.

Moving on to Part B, an examination of the structural model's prediction capabilities was done to investigate the relationship between the CSFs [15]. This evaluation includes a collinearity test (VIF), path coefficient analysis, coefficient of determination (R2), effect size (f2), and predictive relevance (Q2). Based on the results generated through Smart PLS, the reference model of CSFs for improvement initiative fulfilled most of the requirements for the structural model analysis. Figure 5 shows the finalized reference model.

4 Conclusions

In conclusion, this research proposes a reference model of critical success factors derived from the viewpoint of different strategic initiatives to facilitate decision-makers in preparation and assessment when implementing IMI and assist them toward IMI success.

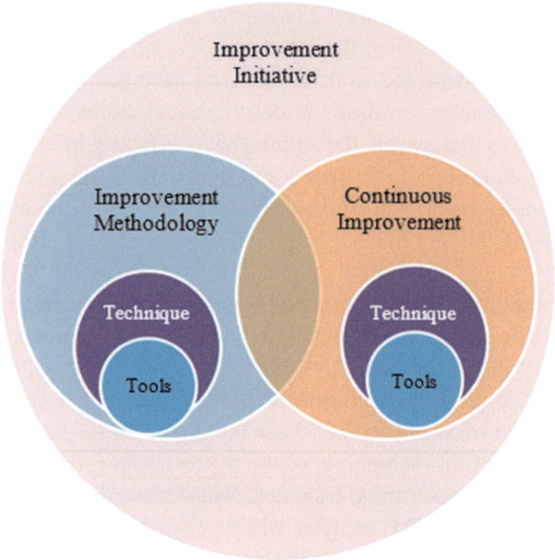

Fig. 1 Level of improvement initiative

For the body of knowledge, the findings from this study provide a holistic view of key factors by providing CSFs and their attributes derived from rigorous literature searching. The key factors integrated into a single CSF model provide IMI practitioners and top management with additional detailed insights regarding CSFs shared from wide-angle perspectives from established researchers studying different initiatives. Thus, it overcomes the lack of the previous models which only focus on CSF for specific initiatives, in which it can be prone to limited understanding and coverage of critical success factors.

In terms of practicality, by having an established reference model, business operators can independently examine which success factors they must prepare and focus on before deploying the improvement initiatives with minimal external help. This will strengthen the effectiveness of improvement activities and help reduce the cost when conducting improvement efforts. The author believes that this study may also encourage decision-makers in viewing other opinions of IMI rather than focusing on specific known initiatives and consider all the key factors before deploying improvement initiative.

Appendixes

See Figs. 1, 2, 3 and 4.

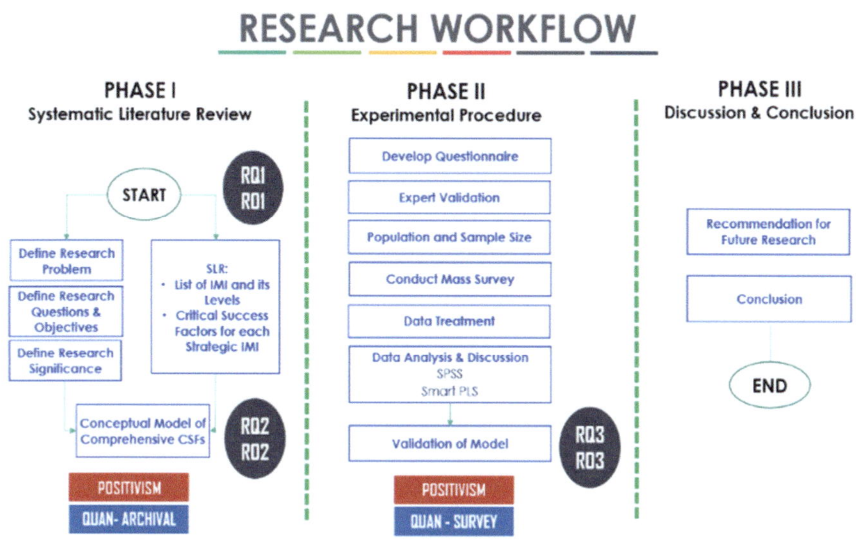

Fig. 2 Research workflow

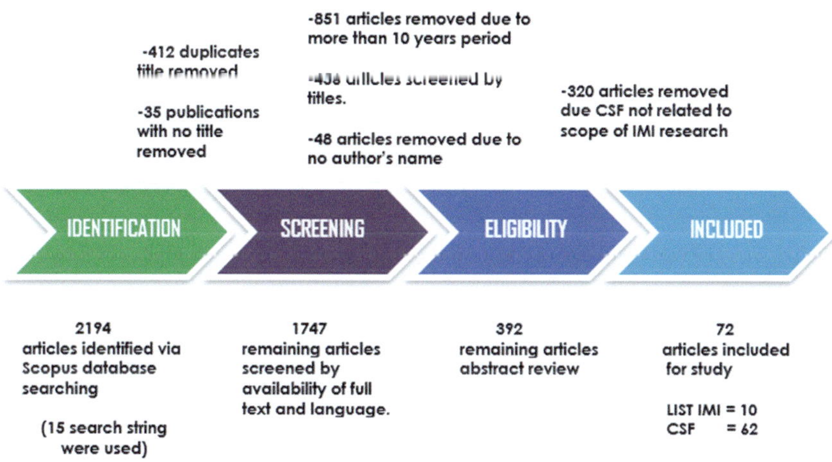

Fig. 3 Number of papers retrieved from SLR

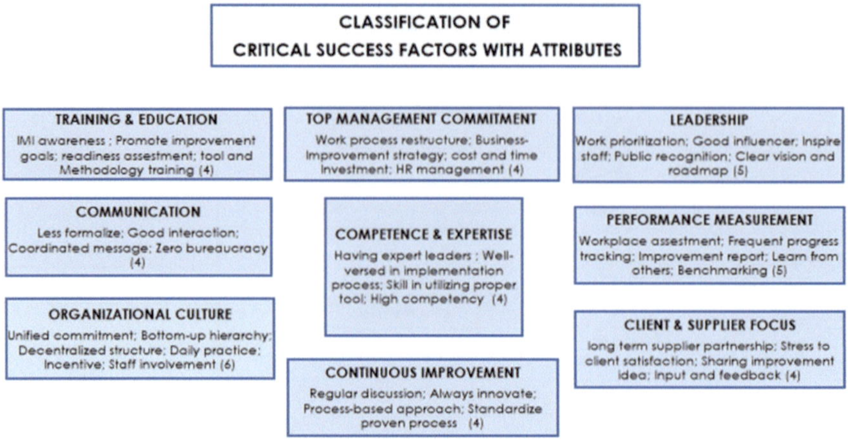

Fig. 4 Conceptual model of critical success factor for strategic initiative

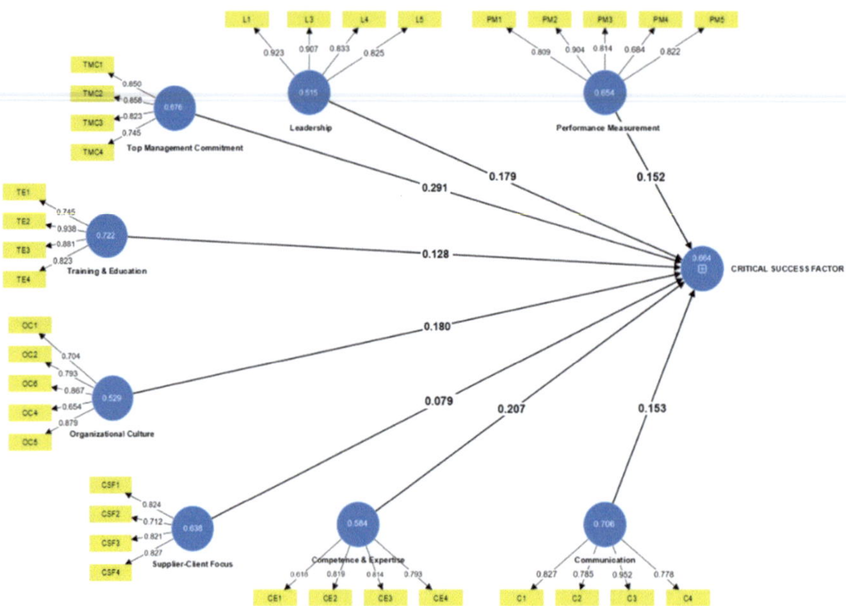

Fig. 5 Finalized critical success factor model for strategic improvement initiative

References

1. S.G.H. Hashemi Nassab, The role of financial management on productivity of the refinery case study: Khangiran Gas Refinery of Iran. **10** (2016). https://doi.org/10.5539/mas.v10n6p44

2. V. Ismyrlis, O. Moschidis, The use of quality management systems, tools, and techniques in ISO 9001:2008 certified companies with multidimensional statistics: the Greek case. Total Qual. Manag. Bus. Excell. **26**, 497–514 (2013)
3. M. Mohammad, R. Mann, N. Grigg, J.P. Wagner, Business excellence model: an overarching framework for managing and aligning multiple organisational improvement initiatives. Total Quality Manag. Bus. Excell. **22**, 1213–1236 (2011)
4. M.I. Abdul Wahab, Development of rational selection model for improvement initiatives. Int. J. Adv. Trends Comput. Sci. Eng. **9**, 174–181 (2020)
5. MPC Annual Report: official Website Malaysia Productivity Corporation. https://www.mpc.gov.my/annual-report
6. M. Sony, J. Antony, S. Naik, How do organizations implement an effective LSS initiative? A qualitative study. Benchmark.: Int. J. **27**, 1657–1681 (2020)
7. D.R. Daniel, Management information crisis. Harvard Business Review (1961), p. 111
8. S.B. El Safty, Critical success factors of lean manufacturing implementation in automotive industry in China. SAE Technical Paper Series (2012)
9. K. Moosa, A. Sajid, Critical analysis of six sigma implementation. Total Qual. Manag. Bus. Excell. **21**, 745–759 (2010)
10. S. Bhasin, An appropriate change strategy for lean success. Manag. Decis. **50**, 439–458 (2012)
11. C.J.F. Cândido, S.P. Santos, Is TQM more difficult to implement than other transformational strategies? Total Qual. Manag. Bus. Excell. **22**, 1139–1164 (2011)
12. J.S. Oakland, S. Tanner, Successful change management. Total Qual. Manag. Bus. Excell. **18**, 1–19 (2007)
13. M.N. Habib, Understanding critical success and failure factors of business process reengineering. Int. Rev. Manag. Bus. Res. (2013)
14. A. Field, Discovering statistics using IBM SPSS statistics: North American edition (2017)
15. W.-L. Shiau, M. Sarstedt, J.F. Hair, Internet research using partial least squares structural equation modelling (PLS-SEM). Internet Res. **29**, 398–406 (2019)

Integration of Quality and Technology Acceptance Model (Q-TAM) to Evaluate Online Teaching and Learning Medium

Noor Syafiqah Redzwan, Sallaudin Hassan, Mohd Haziq Zakaria, Mohd Aliff Afira Sani, Jimisiah Jaafar, Rahmat Roslan Buang, Mohd Farid Shamsudin, and Waqia B. Fokeena

1 Introduction

The unexpected COVID-19 pandemic shocked the world in early 2020. In December 2019, it was first detected in Wuhan, China, and within a few months it has spread to different parts of the world, causing a pandemic [1]. The Ministry of Higher Education (MOHE) declared that teaching and learning need to be conducted online. However, students need to attend physical laboratories, workshops, and design

N. S. Redzwan · S. Hassan (✉) · M. H. Zakaria · M. A. A. Sani · J. Jaafar · R. R. Buang
Malaysian Institute of Industrial Technology, Universiti Kuala Lumpur, Persiaran Sinaran Ilmu, Johor Bahru, Malaysia
e-mail: sallaudin@unikl.edu.my

N. S. Redzwan
e-mail: syafiqahredzwan@gmail.com

M. H. Zakaria
e-mail: mohdhaziq@unikl.edu.my

M. A. A. Sani
e-mail: mohdaliff@unikl.edu.my

J. Jaafar
e-mail: jimiasiah@unikl.edu.my

R. R. Buang
e-mail: rahmatroslan@unikl.edu.my

M. F. Shamsudin
Business School, Universiti Kuala Lumpur, Kuala Lumpur, Malaysia
e-mail: mfarid@unikl.edu.my

W. B. Fokeena
Eduxplora Limited, Port Louis, Mauritius
e-mail: waqia@eduxplora.com

A. Fitriadhy et al. (eds.), *Advancing Interdisciplinary Knowledge and Innovation*, SpringerBriefs in Applied Sciences and Technology,
https://doi.org/10.1007/978-3-032-01309-5_13

studios. Students were also allowed to enter campus to use specialized equipment to carry out their research or project, new students, and Technical and Vocational Training (TVET) education program. Using an online platform, specifically for live classes like Microsoft Teams, requires a broadband connection and a high-speed Internet connection, which cannot always be guaranteed, perhaps from side educators or students. Aside from the issue, a lack of guidance and motivation in dealing with the changes in their learning, spending too much time on a personal computer (PC), and working on digital platforms always carries the risk of data loss due to virus attacks or human error [2].

Several aims have been identified in this research as follows:

i. To measure the effects of Software Product Quality on Perceived Usefulness (PU), Perceived Ease of Use (PEOU), and Attitude Towards Use (ATU) in the adoption of Microsoft Teams for teaching and learning.
ii. To evaluate the effects of Perceived Usefulness (PU) on Perceived Ease of Use (PEOU) and Attitude Towards Use (ATU) in the adoption of Microsoft Teams for teaching and learning.
iii. To evaluate the effects of Perceived Ease of Use (PEOU) on Attitude Towards Use (ATU) in the adoption of Microsoft Teams for teaching and learning.

2 Literature Review

Software Product Quality (ISO/IEC 25010)

The ISO/IEC 25010 is a part of the System and Software Quality Requirement and Evolution (SquaRE) series of the ISO/IEC 25000 to ISO/IEC 25099 entitled Systems and software engineering with Systems and software Quality Requirements and Evaluation. The product quality model is comprised of eight criteria that pertain to both the static and dynamic qualities of software and the computer system. The eight criteria are functional suitability, performance efficiency, compatibility, usability, reliability, security, maintainability and portability. These criteria are able to produce a comprehensive assessment on the quality aspect for technology acceptance.

Perceived Usefulness

Perceived usefulness (PU) is able to influence users' intention to use and accept the new medium of online teaching and learning. A study by [3] mentioned that usefulness is about the good someone trusts that by using a certain method or system that increases the performance. According to [4], a platform like E-Learning will help to improve the study performance. It is also an online course which is able to boost productivity. There is a positive effect between e-learning and improved knowledge and effectiveness of online learning. In contrast, according to [5], students' attitudes have significantly low influence toward students' attitudes in accepting online learning. The main reason is less social networking and physical communication with the instructor or team members.

Perceived Ease of Use

Perceived ease of use (PEOU) is an indicator which influences the user's attitude toward accepting the way an online class is to be conducted. PEOU is referred to when user thinks that the medium has significant benefit to the user. It is aligned with the definition of PEOU where a user believes that the system is free of efforts. In consequence [6], conducted a research to study the acceptance level among the students on e-learning and investigated the antecedence that influences teacher education in Malaysia. The research revealed that convenience is the main factor that influenced student acceptance level. In addition, a study by [7] found that PEOU significantly influence the perceived usefulness.

Attitude Towards Use

Attitude towards use is the level where a person makes assessment and links the system with the career. Attitude also plays a role in order for an individual to accept new technology in the market. The adoption of a person toward a new technology is dependent on the attitude which is influenced by the perceive of use and perceived usefulness. A research by [8] found that most of the English e tutoring is better than a face-to-face approach without the Internet. Thus, the attitude towards use (ATU) was positively affected by PU and PEOU with the behavioral intention to use (BIU) was positively affected by attitude towards using (ATU).

Conceptual Model

TAM does not provide an adequate foundation for analyzing the impact of external variables. The impact that many studies have had on the new expanded technology acceptance model (TAM) to accomplish the research's desired outcomes by integrating external variables into the TAM. One of the variables used in TAM is the quality variable, although other external variables might be used. The previous researchers who were interested in the quality technology acceptance model (Q-TAM) are limited references. Services quality, information flow quality, logistics fulfilment quality, system quality, and product quality are all recorded quality variables that have been implemented in the TAM. The variable from technology

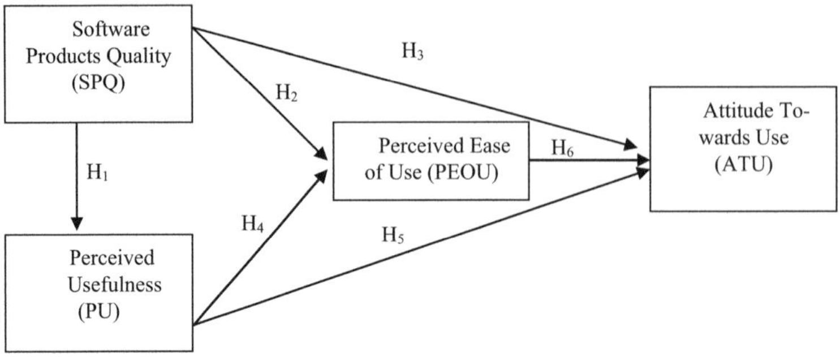

Fig. 1 Q-TAM framework

acceptance model (TAM) and software product quality model from ISO/IEC 25010:2011 was integrated to affect the acceptance adoption of Microsoft Teams as a medium of online teaching and learning in UniKL MITEC's higher TVET. As illustrated in Fig. 1, the original framework for software product quality defined in ISO/IEC 25010:2011 was directly applied as an external variable in the technology acceptance model, which is very similar to the conceptual framework [9]. However, just two variables from TAM were removed from the original framework because they were not necessary. The reasons for the reduction are based on the variable that is present in the original software quality.

3 Methodology

This study applied the deductive methodology. The data analysis is conducted based on the quantitative approach. Past research works also have been referred to as a guideline to select the current method of research and analysis. The questionnaire is the primary data collection used in this research to gather the information from respondents. The questionnaire items involved are the software product quality attributes, perceived usefulness, perceived ease of use, and attitude towards use which are adapted from the model in ISO/IEC 25010:2011 and the technology acceptance model (TAM).

The population whose respondents are among the students at the higher TVET institute is in the undergraduate student of bachelor engineering technology (BET) program at the Universiti Kuala Lumpur Malaysian Institute of Industrial Technology. Students who are required to use hands-on application and implementation of technical expertise to complete engineering tasks are the target audience for this study. Based on [9], at least 217 respondent are required for this research population by setting a confidence interval of 5% to a 95% confidence level. The stratified random sample involves a process of stratification or segregation. The next step is random selection of subjects from each stratum. By using this approach, we can minimize the possibilities of human error and bias in respondent selection.

Once the pilot test analysis is completed, next proceed with the calculation of the for major data collection. First, the data were uploaded into IBM SPSS Statistics 25 to perform a preliminary data analysis. The descriptive analysis focused on frequencies, and percentage common method bias, and exploratory factor analysis. At a later stage, the proposed model was tested using the PLS-SEM technique with the software package Smart-PLS to justify the confirmatory factor analysis (CFA), normality, reliability, validity, and model fit.

4 Results and Discussion

The total number of respondents is 217. Based on the record, 127 are male (58.5%) and 90 female (41.5%). Because this study is focused on students pursuing a degree, the majority of age groups are 23–24 years old (31.3%) and followed by other age groups which are 25 years old and above (31.3%), 21–22 years old (20.3%), and lastly 20 years old and below (11.1%). A high number of respondents says yes (51.6%) to using Microsoft Teams for other purposes except formal teaching and learning and others no (48.4%). Most students are using server's Internet connection from Unifi (23.0%) and U-mobile (21.2%) while using the Microsoft Team software. The majority of students are most likely to use laptops (82.5%) while using software Microsoft Teams for online class, group discussions, or any other purpose.

5 Summary of Findings

Analysis on research question 1 found that there is a direct relationship between Software Product Quality and Perceived Usefulness, Perceived Ease of Use, and Attitude Towards Use. In addition, the direct relationship between every two constructs is significant. From this research question 1 (RO1), three hypotheses related to this question as indicated below:

H1: There is a significant relationship between Software Quality and Perceived Usefulness (PU).
H2: There is a significant relationship between Software Quality and Perceived Ease of Use (PEOU).
H3: There is a significant relationship between Software Quality and Attitude Towards Use (ATU).

By data analysis using Smart-PLS by the bootstrapping procedure, it was found that the path coefficient value for H1 is 0.64, H2 is 0.59, and H3 is 0.25, while the T-score (indicator of significance) for H1 is 14.64, H2 is 7.32, and H3 is 2.59, which is above the threshold of 1.96 (at a 95% confidence level). Based on this result, it is concluded that software product quality has significant and direct effects on perceived usefulness. Software product quality has significant and direct effects on PEOU, and software product quality has significant and direct effects on attitude towards use. Thus, H1, H2, and H3 are supported and RQ1 is addressed.

The second research question focused on the relationship between PU and PEU. Interestingly, the study revealed that there is no significant and direct effect between both constructs. However there is a significant and direct effect relationship between Perceived Usefulness and Attitude Towards Use. From this research question 2 (RO2), two hypotheses related to this question were derived as indicated below:

H1: There is significant relationship between Perceived Usefulness (PU) and Perceived Ease of Use (PEOU).
H2: There is significant relationship between Perceived Usefulness (PU) and Attitude Towards Use (ATU).

By a bootstrapping procedure, it was found that the path coefficient value for H4 is 0.10, and H5 is 0.43, while the T-score (indicator of significance) for H4 is 1.1 and for H5 it is 4.37, which is only H5 above the threshold of 1.96 (at a 95% confidence level). Based on this result, it is concluded that Perceived Usefulness has non-significant and no direct effects on Perceived Ease of Use. But Perceived Usefulness has significant and direct effects on Attitude Towards Use. Thus, only H5 are supported for RQ2. Based on the last research question of this study, it was found that there is a non-significant and no direct effect relationship between Perceived Ease of Use and Attitude Towards Use. From this research question 3 (RO3), a hypothesis related to this question is derived as indicated below:

H1: There is significant relationship between Perceived Ease of Use (PEOU) and Attitude Towards Use (ATU).

Through data analysis using SmartPLS by a bootstrapping procedure, it was found that the path coefficient value for H6 is 0.15. At the same time, the T-score (an indicator of significance) for H5 is 1.87, which is below the threshold of 1.96 (at a 95% confidence level). Thus, it is concluded that Perceived Ease of Use has non-significant and no direct effects on Attitude Towards Use. Thus, H6 are not supported for RQ3.

6 Conclusion

This study uncovered a strong relationship between tested constructs. There is a significant and direct effect of Software Product Quality on Perceived Usefulness. Software Product Quality also has significant and direct effects on Perceived Ease of Use. There are also significant and direct effects of Software Product Quality on Attitude Towards Use. It is also discovered that Perceived Usefulness has significant and direct effects on Attitude Towards Use. However, Perceived Usefulness has non-significant and no direct effects on Perceived Ease of Use. Similarly, Perceived Ease of Use has non-significant and no direct effects on Perceived Ease of Use. Based on this research, Q-TAM model can be applied in assessing the new technology acceptance level by the society. This approach is suitable to be applied in higher education especially in TVET. With the current significant support by the government toward new technology in TVET, more research should be conducted in this area. Future study also can make a comparison between TVET institutional and other categories of universities.

References

1. D. Hui, The continuing 2019-nCoV epidemic threat of novel coronaviruses to global health. J. Infect. Dis. **91**, 264–266 (2020)
2. H.I. Ibrahim, H. Khan, TVET Learning in the time of the pandemic: the Malaysian insight. **6**(2), 1098–1105 (2022)
3. F.D. Davis, Perceived usefulness, perceived ease of use, and user acceptance of information technology. J. Manag. Inf. Syst. **13**(3), 319–339 (1989)
4. D. Pal, V. Vanijja, Perceived usability evaluation of Microsoft Teams as an online learning platform during COVID-19 using system usability scale and technology acceptance model in India. Child Youth Serv. Rev. **119**, 105–109 (2020)
5. S.N.S. Allam, M.S. Hassan, R.S. Mohideen, A.F. Ramlan, R.M. Kamal, Online distance learning readiness during covid-19 outbreak among undergraduate students. Int. J. Acad. Res. **10**(5), 642–657 (2020)
6. M. Suhaimi Taat, A. Francis, Factors influencing the students' acceptance of e-learning at teacher education institute: an exploratory study in Malaysia. Int. J. High. Educ. **9**(1), 133–141 (2020)
7. F.D. Davis, User acceptance of information technology: system characteristics, user perceptions and behavioral impacts. Int. J. Man-Mach. Stud. **38**(3), 475–487 (1993)
8. K.G. Boakye, T. McGinnis, V.R. Prybutok, Q-TAM-a quality technology acceptance model for technology operations managers. J. Oper. Manag. **7**(2), 13–23 (2014)

The Influence of Green Port Practices on Corporate Image: A Malaysian Perspective

Amirul Husaini Ibrahim, Amayrol Zakaria, Aminuddin Md. Arof, and Shamila Azman

1 Introduction

In recent years, there has been an increasing emphasis on environmental sustainability and corporate social responsibility in various industries, including the maritime sector. Ports, as crucial components of the global supply chain, are also striving to adopt green initiatives to reduce their environmental impact and enhance their corporate image.

Green port practices encompass a range of strategies and initiatives aimed at minimizing the environmental footprint of port operations. These practices include the adoption of eco-friendly technologies, energy-efficient operations, waste management strategies, and the promotion of environmental awareness among port stakeholders [1]. The integration of sustainable practices within port operations is crucial for mitigating environmental risks and enhancing corporate reputation [2]. From Malaysian experience, Arof et al. highlighted that the goal of green port is to reduce or eliminate environmental impacts while increasing port efficiency [3].

A. H. Ibrahim · A. Zakaria (✉) · A. Md. Arof
Malaysian Institute of Marine Engineering Technology, Universiti Kuala Lumpur, Lumut, Malaysia
e-mail: amayrol.zakaria@unikl.edu.my

A. H. Ibrahim
e-mail: amirulhusaini320@gmail.com

A. Md. Arof
e-mail: aminuddin@unikl.edu.my

S. Azman
Department of Water and Environmental Engineering, School of Civil Engineering, Faculty of Engineering, Universiti Teknologi Malaysia, Skudai, Johor, Malaysia
e-mail: shamila@utm.my

A. Fitriadhy et al. (eds.), *Advancing Interdisciplinary Knowledge and Innovation*, SpringerBriefs in Applied Sciences and Technology,
https://doi.org/10.1007/978-3-032-01309-5_14

Impact on Corporate Image

Corporate image refers to the overall perception and reputation of a company among its stakeholders. Green port practices have the potential to positively influence corporate image by demonstrating a commitment to environmental sustainability and social responsibility [4]. Research suggests that environmentally responsible practices can enhance a port's image as an eco-friendly organization, leading to increased stakeholder trust, customer loyalty, and competitive advantage [5]. The successful implementation of the green port practices is influenced by various factors. Government regulations and policies play a significant role in driving sustainability initiatives within the port industry [6]. Collaborative efforts between port authorities, stakeholders, and industry associations are essential for creating a conducive environment for sustainable practices [7]. Adequate financial resources, both from public and private sectors, are crucial for investing in green technologies and infrastructure [8]. Furthermore, employee engagement and training programs are necessary for fostering a culture of sustainability within port organizations [7].

Malaysian Port Industry

The Malaysian port industry has made notable progress in adopting green port practices. For instance, Port Klang, one of the major ports in Malaysia, has implemented various sustainability initiatives, including the use of shore power to reduce vessel emissions and investments in renewable energy sources [9]. These initiatives have not only improved environmental performance but have also positively impacted the corporate image of Port Klang, positioning it as a sustainable and responsible port [10].

2 Methodology

In this study, information from respondents is gathered using the Delphi approach. The questionnaire was distributed to respondents who are knowledgeable about the impact that green port practices have on the perception of businesses in Malaysia's port. Initially, this study employed Cronbach's alpha to evaluate the internal consistency and reliability of survey or test items that were created to measure the same construct. The mean value, median, and standard deviation are determined for the statistical analysis's use in determining the consensus of the experts by using descriptive analysis. This research had been conducted at the container ports in Malaysia and focused on the gathering of information data in selected samples from the population in Malaysia container ports. Based on the Krejcie and Morgan table, this study utilized the purposive sampling method where the total of 150 expert respondents from ports top management, shipping companies, logistics providers, government agency, government enforcement body, and maritime research organization which hold an adequate experience were chosen to respond to the questionnaire. It is supported by the authors of [11] where the study also utilized the Krejcie and Morgan table to determine the appropriate number of respondent studies that focus on the green economy approach.

3 Results and Discussion

Result of R1 Delphi

Table 1 shows the result of the pilot test that has been utilized by Cronbach's alpha. Cronbach's alpha value is 0.934 which means that the set of questionnaires are reliable to utilize for the gathering process. The reliability statistic is shown in Table 1.

Based on Table 2, the highest mean value is "energy-efficient operations" at which the mean value is 6.68 followed by "waste management and recycling" with the mean value being 6.66, "environmental monitoring and reporting" with the mean value being 6.57, "eco-friendly cargo handling equipment" with the mean value being 6.54, "shore power facilities and infrastructure" with the mean value being 6.52, "renewable energy adoption" with the mean value being 6.41, "collaboration and stakeholder engagement" with the mean value being 6.35, "environmental training and education" with the mean value being 6.30, "water conservation" with the mean value being 6.28, and the last is "green building design" with the mean value being 6.24. Basically, the data are close to the mean value or average of the data set, as shown by the fact that all variables have standard deviations less than 1.0. Since all of the variables fall within the range of 6–7 on the Likert scale (agree to strongly agree), it is obvious that all of the variables or practices are currently being used in the organization. However, the expert respondent made no recommendations to include any new variables that would be appropriate for the study. As a result, the R2 questionnaire and the findings were given to each respondent separately with the intention of reaching an expert consensus.

Table 1 Reliability statistics

Cronbach's alpha	No of items
0.934	20

Table 2 First round survey

Determinant	N	Mean	S.D.
Energy-efficient operations	150	6.68	0.307
Waste management and recycling	150	6.66	0.376
Environmental monitoring and reporting	150	6.57	0.347
Eco-friendly cargo handling equipment	150	6.54	0.350
Shore power facilities	150	6.52	0.327
Renewable energy adoption	150	6.41	0.357
Collaboration and stakeholder engagement	150	6.35	0.476
Environmental training and education	150	6.30	0.327
Water conservation	150	6.28	0.327
Green building design	150	6.24	0.357
Valid N (listwise)	150		

Result of R2 Delphi

Based on Table 3, the highest mean value is "energy-efficient operations" at which the mean value is 6.58 followed by "waste management and recycling" with the mean value being 6.56, "environmental monitoring and reporting" with the mean value being 6.47, "eco-friendly cargo handling equipment" with the mean value being 6.43, "shore power facilities and infrastructure" with the mean value being 6.41, "renewable energy adoption" with the mean value being 6.40, "collaboration and stakeholder engagement" with the mean value being 6.34, "environmental training and education" with the mean value being 6.29, "water conservation" with the mean value being 6.26, and the last is "green building design" with the mean value being 6.22. As for the standard deviation, all variables are less than 1.0 which indicated that the data are close to the mean value or average of the data set. Obviously, since all the variables are under the 6–7-point Likert scale (agree to strongly agree), it can be indicated that all the expert respondents strongly agree that the practices are currently in practice in the organization. Thus, the finding revealed that there is consensus judgement among the expert respondents since the median is 6 and standard deviation is less than 1.0. Consequently, reaching expert consensus can be done without needing to participate in the following Delphi round. Thus, the next section is the process to determine the relationship between green practices and corporate image.

Pearson Correlation Analysis

Table 4 shows the correlation between two variables, one of which is the dependent variable and the other is the independent variable. The "waste management and

Table 3 Second round survey

Determinant	N	Mean	S.D.
Energy-efficient operations	150	6.58	0.317
Waste management and recycling	150	6.56	0.386
Environmental monitoring and reporting	150	6.47	0.357
Eco-friendly cargo handling equipment	150	6.43	0.360
Shore power facilities	150	6.41	0.337
Renewable energy adoption	150	6.40	0.367
Collaboration and stakeholder engagement	150	6.34	0.486
Environmental training and education	150	6.29	0.337
Water conservation	150	6.26	0.357
Green building design	150	6.22	0.360
Valid N (listwise)	150		

recycling" is the highest value compared to another variable which is 0.500, and the data is close to 1 where the result demonstrates that there is a significant relationship between "waste management and recycling" and corporate image. On the other hand, this result also indicated a positive value for all green practices or variables that signify a significant relationship with the image. Obviously, each of the green practices examined above contributes significantly to the improvement of the port's corporate image.

Table 4 Pearson correlation analysis

	1	2	3	4	5	6	7	8	9	10	11
1	1.00	0.50	0.42	0.48	0.46	0.48	0.43	0.44	0.46	0.46	0.46
2	0.50	1.00	0.62	0.00	–0.26	0.24	–0.50	–0.37	–0.37	–0.26	–0.37
3	0.42	0.62	1.00	0.00	–0.19	0.88	–0.37	–0.06	–0.06	–0.19	–0.06
4	0.48	0.00	0.00	1.00	0.30	0.00	–0.59	0.00	0.00	0.30	0.00
5	0.46	–0.26	–0.19	0.30	1.00	–0.13	–0.21	0.57	0.57	0.46	0.57
6	0.48	0.24	0.88	0.00	–0.13	1.00	–0.26	0.26	0.26	–0.26	0.26
7	0.43	0.50	–0.37	–0.59	–0.20	–0.26	1.00	–0.30	–0.30	–0.19	–0.30
8	0.44	–0.17	–0.06	0.00	0.57	0.26	–0.30	1.00	0.50	0.30	0.50
9	0.46	–0.26	–0.19	0.30	0.30	–0.13	–0.21	0.57	1.00	0.46	0.46
10	0.46	–0.26	–0.19	0.30	0.30	–0.13	–0.21	0.57	0.46	1.00	–0.37
11	0.48	0.24	0.88	0.00	–0.13	0.48	0.24	0.88	0.00	–0.13	1.00

4 Discussion

In retrospect, there are ten green practices that have been revealed in this study. Based on the expert consensus from this study, the proper green practices in container port in Malaysia are as listed below.

i. Energy-efficient operations
ii. Waste management and recycling
iii. Environmental monitoring and reporting
iv. Eco-friendly cargo handling equipment
v. Shore power facilities and infrastructure
vi. Renewable energy adoption
vii. Collaboration and stakeholder engagement
viii. Environmental training and education
ix. Water conservation
x. Green building design.

It is important to note that these green practices are based on the ratings provided and may vary depending on geographical, specific contexts, and priorities. However, the ratings can serve as a starting point for understanding the perceived importance of these green port practices in relation to corporate image and sustainability development. Therefore, it is essential to conducting a comprehensive assessment of each port's specific needs, available resources, and local regulations to tailor the green practices accordingly. Additionally, considering the evolving nature of sustainability practices, it's recommended to stay updated with the latest advancements and international best practices in the field.

5 Conclusion

In conclusion, the relationship between green port practices and corporate image is intertwined and mutually reinforcing. Green port practices, encompassing environmentally responsible initiatives and sustainable operations, have a significant impact on a port's corporate image and contribute to broader sustainability development goals. By adopting green port practices, ports demonstrate their commitment to reducing environmental impacts, promoting resource efficiency, and mitigating climate change. These practices not only enhance the environmental performance of ports but also contribute to the overall sustainability of the maritime industry and the communities they serve. Such proactive efforts can bolster a port's corporate image as a responsible and forward-thinking organization that prioritizes environmental stewardship. A positive corporate image, in turn, can lead to several benefits for ports. It can attract environmentally conscious customers, including shipping lines, logistics providers, and cargo owners, who seek to align their operations with sustainable partners. This, in turn, can increase port competitiveness and attract new business opportunities. Additionally, a strong corporate image as a green port can enhance relationships with regulators, local communities, and other stakeholders, fostering collaboration and support for sustainable initiatives.

Moreover, the adoption of green port practices contributes to broader sustainability development. It aligns with global and national sustainability agendas, such as the United Nations Sustainable Development Goals (SDGs) and national environmental targets. Green port practices can reduce carbon emissions, conserve resources, protect biodiversity, and improve air and water quality, thus positively impacting the ecological and social aspects of sustainability. Eventually, effective green port practices not only improve the environmental performance of ports but also enhance their corporate image and contribute to sustainability development. By integrating sustainable principles into their operations, ports can strengthen their competitiveness, attract environmentally conscious partners, and contribute to global efforts in achieving a more sustainable and resilient future.

References

1. K. Bichou, R. Gray, A logistics and supply chain management approach to port performance measurement. MPM **39**(2), 171–189 (2012)
2. Y. Cheng, J. Smith, A. Johnson, The integration of sustainable practices within port operations: Mitigating environmental risks and enhancing corporate reputation. IMS **15**(3), 45–62 (2017)
3. A. M. Arof, A. Zakaria, and N. S. F. A. Rahman, Green port indicators: A Review. in *AEPT II* (2021), pp. 237–256
4. S. Kuo, C. Lee, H. Chang, The impact of green port practices on corporate image: A study on environmental sustainability and social responsibility. Int. J. Sustain. Transp. **27**(2), 89–105 (2019)
5. M.V. Russo, P.A. Fouts, Enhancing stakeholder trust, customer loyalty, and competitive advantage through environmentally responsible practices: A study on port organizations. Sustain. Bus. Int. J. **12**(3), 45–61 (1997)
6. S. Saurí, P. Du, The role of government regulations and policies in driving sustainability initiatives within the port industry. JPR **25**(2), 45–62 (2019)
7. T.J. Lam, H. Li, Collaborative efforts among port authorities, stakeholders, and industry associations for fostering a conducive environment for sustainable practices. IJSP **17**(3), 123–140 (2020)
8. T. Notteboom, J.P. Rodrigue, The importance of adequate financial resources from both public and private sectors for investing in green technologies and infrastructure within the port industry. JST **32**(4), 78–95 (2018)
9. C. Lam, R. Tan, S. Lee, Sustainability initiatives at Port Klang: A case study of reducing vessel emissions through shore power and investments in renewable energy sources. J Sustain Ports **38**(2), 145–162 (2021)
10. A. Koh, L. Tan, C. Wong, The impact of sustainability initiatives on the environmental performance and corporate image of Port Klang: A case study. JSP **25**(4), 78–95 (2020)
11. P. Ataei, H. Karimi, S. Moradhaseli, M.H. Babaei, Analysis of farmers' environmental sustainability behavior: the use of norm activation theory (a sample from Iran). AJGS **15**(9), 859 (2022)

Challenges Faced by Seaports in Adoption of Industrial Revolution 4.0: A Review

Kaannappar Ponniah Chandrasekaran, Amayrol Zakaria, and Aminuddin Md. Arof

1 Introduction

Logistics and supply chain management play a crucial role in reducing expenses and gaining a competitive advantage for organizations. Maritime transport is a key component of the global supply chain, with 90% of goods transported across geographical boundaries relying on maritime transport [1]. Seaports are vital economic hubs, influenced by various factors including socioeconomic changes, globalization, and technological advancements [2]. Effective stakeholder interaction is essential for resource management and operational optimization in seaports [3].

Seaports serve as major economic centers and are responsible for managing and promoting domestic and international trade on a large scale [4]. They offer diverse services, including goods delivery, storage, consolidation, inventory management, and more [4]. The location of seaports is influenced by cargo types, demand for various items, the need for constant dredging to accommodate larger vessels, and significant infrastructure requirements [5].

The emergence of Industry 4.0 (IR4.0), driven by the integration of information and communication technology (ICT), is transforming seaports into smart ports capable of optimizing their operations. Smart ports use ICT to enhance shipping and port operations, improving economic competitiveness and sustainability [6].

K. P. Chandrasekaran · A. Zakaria (✉) · A. Md. Arof
Malaysian Institute of Marine Engineering Technology, Universiti Kuala Lumpur, Lumut, Malaysia
e-mail: amayrol.zakaria@unikl.edu.my

K. P. Chandrasekaran
e-mail: kaannappar.chandrasekaran@s.unikl.edu.my

A. Md. Arof
e-mail: aminuddin@unikl.edu.my

A. Fitriadhy et al. (eds.), *Advancing Interdisciplinary Knowledge and Innovation*, SpringerBriefs in Applied Sciences and Technology,
https://doi.org/10.1007/978-3-032-01309-5_15

This technology allows for the efficient management of traffic operations, monitoring conditions, and connecting with vehicles and centers [7]. IR 4.0 incorporates advancements such as robotics, big data, cloud manufacturing, and augmented reality, impacting operational processes and global supply chains [8].

The adoption of IR 4.0 technologies in seaports can lead to benefits such as improved data access, machinery maintenance, and cost savings [9]. However, challenges include the need for significant investments, a shortage of trained personnel, and the readiness of the seaport sector for such transformations [7, 10].

Ultimately, IR 4.0 is reshaping seaports into smarter, more efficient, and environmentally friendly hubs of the global supply chain, with the potential to enhance competitiveness and foster industry growth [10]. However, these transformations are not without their difficulties, as the adoption of advanced technologies and the readiness of seaports vary across nations. The concept of "smart port development" is a progressive undertaking aimed at integrating cutting-edge technology into port operations for enhanced performance [10].

2 Literature Review

Seaport

A port, as described by [11], is a location with berthing and mooring facilities for boats and vessels, facilitating the movement of people and cargo between ships and the shore. There are various types of ports, including dry ports, warm-water ports, inland ports, fishing ports, and seaports. Marine transportation is the primary mode for transferring cargo, accounting for 90% of the shipments.

Seaports encompass conventional terminals (e.g., break bulk, dry bulk, liquid bulk) and container terminals, categorized into civil engineering features, administrative functions, and operational functions, as per [12]. A port includes both sea-to-land passageways with berthing facilities and connections to rail and road networks, often with nearby industrial areas.

Industrial Revolution

The industrial revolution started in the late eighteenth century and progressed through several phases. The first industrial revolution introduced mechanization with steam and waterpower. The second revolution harnessed electric power for mass production, and the third brought automation through electronic and information technologies [13].

Building on the Third Industrial Revolution, the Fourth Industrial Revolution is emerging, characterized by the convergence of physical, digital, and biological realms. The main goal of IR 4.0 is to enhance data availability and precision using technologies like IoT, smart sensors, advanced robotics, and more. It relies on the latest telecommunications networks like 5G [14].

Maritime 4.0 refers to IR 4.0 applied to the maritime sector. It focuses on digitalization, automation, and information-oriented industries. Digitalization involves transferring manual processes to digital environments to streamline operations and manage vast, heterogeneous data efficiently. Technologies like blockchain and digital twins are introduced to accelerate port management and operations, enhancing output rates and contributing to the national economy [10, 15, 16]. This concept is being implemented in ports worldwide, with countries like Malaysia embracing it as part of their development [10].

Factors Influencing Implementation

The successful implementation of IR 4.0 depends on several key factors. Even though the Government regulations and policies are pivotal in driving sustainability initiatives in the port industry, collaboration among port authorities, stakeholders, and industry associations is essential for creating an environment conducive to sustainable practices [18]. Adequate financial resources, from both the public and private sectors are crucial for investing in green technologies and infrastructure. Additionally, fostering a culture of sustainability within port organizations requires employee engagement and training programs.

This review paper conducts an extensive analysis of existing literature to focus on the study's key aspects. It references various research articles from open literature. The study examines the challenges associated with implementing IR 4.0 and explores the connection between these challenges and the role of policymakers in addressing them. It systematically presents the theoretical framework, empirical evidence, and identifies the research gap that this study aims to fill.

3 Discussion

The articles discuss various challenges related to infrastructure, logistics, and human resources in the countries of Vietnam, Malaysia, and Egypt. According to [17], Vietnam faces issues with its sparse and underdeveloped logistics and transportation infrastructure, particularly in the railway sector. This hampers its competitiveness in the international market. Malaysia, as noted by [7], struggles with capacity limitations in its rail transport, single-track systems, limited service numbers, and other constraints [10]. The author of [10] also revealed that Egypt's lack of infrastructure is due to the substantial financial requirements for upgrading port facilities, impacting the development of smart seaports aligned with Industry 4.0.

All three countries also face a shortage of qualified maritime experts, hindering their logistics and port operations. Skilled human resources are essential for the success of any industry. Egypt, Malaysia, and Vietnam lack competent personnel with skills, knowledge, and proficiency in specialized English for international seaport operations. Economic and political factors play a role in the adoption of IR 4.0. Government funding and budget issues pose challenges in Egypt and Malaysia, affecting infrastructure development and IR 4.0 implementation. Malaysia faces uncertainty regarding the return on investment in IR 4.0 due to high costs.

Data sharing is another common challenge. Companies are reluctant to open their systems to the public for fear of competitiveness loss. Collaboration and networking should be considered before implementing IR 4.0, as data sharing plays a crucial role in decision-making, equipment projection, and facility renovations. One major concern is the lack of integration and data exchange among various departments in seaports, leading to inefficiencies and maintenance issues. The authors have made significant efforts to present their data analysis findings, but they could benefit from reducing textual complexity. For instance, [7] highlight the importance of trust among stakeholders for successful IR 4.0 implementation in Malaysia.

Briefly, the challenges faced by Vietnam, Malaysia, and Egypt in infrastructure development, human resources, economic and political factors, and data sharing hinder the successful adoption of IR 4.0 in their respective seaport operations. To overcome these challenges, these countries need to address infrastructure issues, invest in skilled human resources, and promote data sharing and integration in their seaport systems.

4 Conclusion

In retrospect, there are 9 key challenges faced by seaports in embracing IR 4.0 namely.

i. Infrastructure and Connectivity
ii. Skill Gap and Workforce Readiness
iii. Data Integration and Interoperability
iv. Cybersecurity and Data Privacy
v Cost and Return on Investment
vi Regulatory and Legal Framework
vii Collaboration and Stakeholder Engagement
viii Acceptance and Cultural Change
ix Environmental Sustainability.

The emergence of IR 4.0 has introduced significant technological advancements to seaports, promising improved operations and sustainability. However, integrating these technologies faces several key challenges. Infrastructure and connectivity upgrades are essential, requiring enhanced network capabilities and seamless communication. Skill gaps and workforce readiness are critical, demanding training for employees to operate advanced technologies. Data integration and interoperability are pivotal for effective decision-making, and cybersecurity becomes paramount in protecting critical infrastructure.

Moreover, the economic aspect is crucial, with seaports needing to assess costs and ROI meticulously. Regulatory and legal frameworks pose obstacles, necessitating compliance with changing regulations. Stakeholder collaboration and integration are complex, as diverse stakeholders must align interests and share data. Overcoming resistance to cultural change and fostering innovation is essential, and environmental sustainability remains a growing concern. Addressing these challenges is key to the successful adoption of IR 4.0 technologies in seaports.

The adoption of IR 4.0 technologies presents numerous challenges for seaports seeking to enhance their efficiency, productivity, and competitiveness. Throughout this review, we have identified and analyzed key challenges faced by seaports in the adoption of these transformative technologies.

Addressing these challenges requires a holistic and collaborative approach. Seaports must actively engage with stakeholders, including port authorities, terminal operators, shipping lines, and logistics providers, to align interests, share data, and establish effective communication channels. Additionally, close collaboration between seaports and regulatory bodies is vital to create an enabling environment that supports innovation while ensuring compliance with relevant regulations and legal frameworks.

Seaports should also invest in continuous workforce development, providing comprehensive training and upskilling opportunities to ensure employees are equipped to operate and maintain advanced technologies. Embracing a culture of innovation and change will foster acceptance and facilitate the adoption of IR 4.0 technologies.

In conclusion, while the adoption of IR 4.0 technologies presents challenges, seaports have the opportunity to enhance their operations, competitiveness, and sustainability. By addressing the identified challenges and implementing effective strategies, seaports can leverage these technologies to transform their processes, optimize resource utilization, and drive future growth in the maritime industry.

Acknowledgment The authors are appreciative of the Malaysian Institute of Marine Engineering Technology at the University of Kuala Lumpur for creating a conducive environment for conducting this review.

References

1. R. Zhang, M. Roe, *Maritime container port security: USA and European perspectives* (Palgrave Macmillan; Springer International Publishing, Basel, Switzerland, 2018)
2. Hlali, S. Hammami, Seaport concept and services characteristics: Theoretical test. Open Transp. J., **11**(1), 120–129 (2017). [Online]. https://doi.org/10.2174/1874447801711010120
3. European intermodal transportation. MABR, **4**(2), 151–168 (2019). [Online]. https://doi.org/10.1108/mabr-09-2018-0038

4. A. Montwiłł, The role of seaports as logistics centers in the modeling of the sustainable system for distribution of goods in urban areas. Procedia Soc. Behav. Sci. **151**, 257–265 (2014). https://doi.org/10.1016/j.sbspro.2014.10.024
5. A. Olusegun Onifade, New seaport development-prospects and challenges: Perspectives from Apapa and Calabar Seaports, Nigeria. Logistics, **4**(2), 8 (2020). https://doi.org/10.3390/logistics4020008
6. K.L.A. Yau, S. Peng, J. Qadir, Y.C. Low, M.H. Ling, Towards smart port infrastructures: Enhancing port activities using information and communications technology. IEEE Access **8**, 83387–83404 (2020). https://doi.org/10.1109/ACCESS.2020.2990961
7. N.H. Mohd Salleh, M. Selvaduray, J. Jeevan, A.H. Ngah, S. Zailani, Adaptation of industrial revolution 4.0 in a seaport system. Sustainability, **13**, 10667 (2021). https://doi.org/10.3390/su131910667
8. A. Pereira, F. Romero, A review of the meanings and the implications of the Industry 4.0 concept. Procedia Manuf. **13**(3), 1206–1214 (2017)
9. L. Heilig, E. Lalla-Ruiz, S. Voß, Port-IO: An integrative mobile cloud platform for real-time inter-terminal truck routing optimization. Flex. Serv. Manuf. J. 504–534 (2017). [Online]. https://doi.org/10.1007/s10696-017-9280-z
10. S. Othman, S. El Gazzar, M. Knez, Investigating the influences of smart port practices and technology employment on port sustainable performance: The Egypt Case. Sustainability, **14**, (2022). [Online]. https://doi.org/10.3390/su142114014
11. S. Islam, T.L. Olsen, Factors affecting seaport capacity, (2015)
12. P.M. Alderton, G. Saieva, Port management and operations. Taylor & Francis, (2013)
13. M. Xu, J.M. David, S.H. Kim, The fourth industrial revolution: opportunities and challenges, vol. 9, no. 2, pp. 90–95 (2018). [Online]. https://doi.org/10.5430/ijfr.v9n2p90
14. P. Sullivan, S. Desai, J. Sole, M. Rossi, L. Ramundo, S. Terzi, Maritime 4.0–Opportunities in digitalization and advanced manufacturing for vessel development. Procedia Manuf., **42**, 246–253 (2020). [Online]. https://doi.org/10.1016/j.promfg.2020.02.078
15. O.P. Brunila, V.K. Hyrkki, T. Inkinen, Hindrances in port digitalization ? Identifying problems in adoption and implementation. Eur. Transp. Res. Rev, (2021). [Online]. https://doi.org/10.1186/s12544-021-00523-0
16. A. Shahroom, N. Hussin, A. A. Shahroom, and N. Hussin, Industrial revolution 4.0 and education, vol. 8, no. 9, pp. 314–319 (2018). [Online]. https://doi.org/10.6007/IJARBSS/v8-i9/4593
17. H.P. Nguyen, Sustainable development of logistics in Vietnam in the period 2020-2025. Int. J. Innov. Creativity Chang., (2020)
18. S. Saurí, P. Du, The role of government regulations and policies in driving sustainability initiatives within the port industry. J. Port Res. **25**(2), 45–62 (2019)

MIX
Papier aus verantwortungsvollen Quellen
Paper from responsible sources
FSC® C105338
FSC
www.fsc.org

If you have any concerns about our products,
you can contact us on
ProductSafety@springernature.com

In case Publisher is established outside the EU,
the EU authorized representative is:
Springer Nature Customer Service Center GmbH
Europaplatz 3, 69115 Heidelberg, Germany

Printed by Libri Plureos GmbH
in Hamburg, Germany